PHOBIAS

Helen Saul is a freelance science journalist, contributing to *New Scientist* and reporting on health programmes on BBC Radio Four. She is also news editor of the *European Journal of Cancer*. She is married with two young children and lives in Oxford. She has never had a phobia.

**FOR REFERENCE ONLY**

# PHOBIAS

Fighting the Fear

HELEN SAUL

HarperCollins*Publishers*

The Author and Publishers would like to express their gratitude to the Provost and Scholars of King's College, Cambridge and the Society of Authors as the Literary Representatives of the E. M. Forster Estate and to Random House, Inc. for permission to quote from *Howard's End* by E. M. Forster. The extract from *Nineteen Eighty-Four* by George Orwell (copyright © George Orwell 1949) reproduced by permission of A. M. Heath & Co Ltd on behalf of Bill Hamilton as the Literary Executor of the Estate of the Late Sonia Brownell Orwell and Martin Secker and Warburg Ltd.

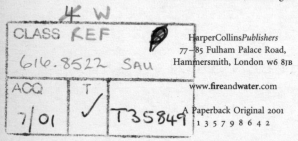

HarperCollins*Publishers*
77–85 Fulham Palace Road,
Hammersmith, London w6 8jb

www.**fire**and**water**.com

A Paperback Original 2001
1 3 5 7 9 8 6 4 2

Copyright © Helen Saul 2001

Helen Saul asserts the moral right to
be identified as the author of this work

A catalogue record for this book is
available from the British Library

ISBN 0 00 638431 5

Set in Minion by
Rowland Phototypesetting Ltd,
Bury St Edmunds, Suffolk

Printed and bound in Great Britain by
Clays Ltd, St Ives plc

*to Mark*

# CONTENTS

# ACKNOWLEDGEMENTS

I am extremely grateful for all the help I received while trying to find a way through the sometimes baffling and conflicting research on phobias. Many doctors and scientists agreed to be interviewed for this book, and I would especially like to mention Randolph Nesse, Isaac Marks, Paul Salkovskis, Graham Davey, Jerome Kagan, Kenneth Kendler, Jonathan Flint, Abby Fyer, Mark George, Don Klein, Laszlo Papp, Tom Uhde, David Barlow, Susan Shaw, Jerry Rosenbaum, Carol Nadelson and Harold Levinson. All were generous with their time and encouragement. I am indebted also to Claire Dimond and Tim Lebon for their helpful comments on the manuscript. For whatever is still wrong with it, the responsibility is, of course, my own. I would like to thank Philip Gwyn Jones at HarperCollins who first saw the potential for a book, and my editor, Georgina Laycock, whose many suggestions have enlivened the text beyond recognition. My love and gratitude go to my family for their patience and support throughout.

Above all, though, I would like to thank so many with phobias who shared their painful and private experiences and brought home to me the real importance of the research. If anyone with such a story to tell is in any way helped by the following pages, the whole project will have been worthwhile.

# Preface

While working on this book, I was often asked which phobias I was writing about. The question initially puzzled me – my intention was always to write about all phobias – though it does perhaps reflect the common but unhelpful assumption that each phobia is a distinct problem with a distinct cause and treatment. In fact, the root causes of fear could apply equally to triskaidekaphobia (fear of the number 13) or gephyrophobia (crossing bridges). This book is intended to be less a self-help book than an exploration of the ideas and thoughts driving progress in the laboratory and the clinic. Everyone with a phobia knows what it is to fight fear and I only hope this book properly acknowledges their courage. But it aims also to focus on the way doctors and scientists are improving our understanding and fighting fear on their behalf.

I am a freelance science and medical journalist and have worked variously in TV, radio, newspapers and magazines. I have no specialist training in phobias, other than a degree in medical sciences. I have never had a phobia or received treatment. My involvement with phobias began in the autumn of 1993 while I was working at *New Scientist*. The then features editor, Bill O'Neill, received a dossier on the subject from a remarkably well-informed former agoraphobic, Mary Dwarka (whose story is told fully in chapter 9). She had researched her own condition in huge depth and was suggesting that *New Scientist* run a feature on agoraphobia. Bill was somewhat perplexed by the range of science and arguments in Mary's work and asked me to look through it to see whether I thought there was enough in the subject to make a feature.

There was more than enough. Enough for a book, I later discovered. At the time, I spoke to a couple of people with experience of phobias and a variety of scientists working on the subject, some suggested by Mary. The resulting feature was optimistic in tone, concentrating on the great advances in our understanding and how they are already translating into new treatments in the clinic.

It sometimes seems that the more one knows about phobias, the less clear everything becomes. Ask members of the public what a phobia is and they will tell you about extreme reactions to spiders or heights. Many will reel off an anecdote about someone they have known with a bizarre fear. And, of course, these are phobias. But as one delves further, boundaries seem to disappear. Where does normal protective fear turn into a phobia? What is the link between generalised anxiety and phobias? Or between panic disorder and panic attacks and phobias? When is agoraphobia a true phobia and when a consequence or the cause of depression? What about avoidance disorders, alcoholism, some personality disorders? Where does phobia end and psychiatry begin? Entire libraries could be devoted to the subject.

Equally, when you ask specialists, they very often produce good clear answers. But ask four professionals in different disciplines and you will be lucky to hear the same thing twice. The amorphous mass of phobia research needs to be simplified for us to get anywhere. Specialists emphasise different aspects, and while none is entirely wrong, none is entirely right either.

This book is an attempt to represent many different points of view. I have chosen to focus a chapter on each approach. Readers may find that some strike a chord while others seem less relevant. Where previous books on phobias have focused on a single approach, this attempts to be an unbiased account of all the advances in all the major schools of thought dealing with phobias.

# Introduction

## *Fight the Fear*

John gazed out of the window in private joy. Normally reticent, he grinned broadly at the beauty of the scene. A perfectly round orange sun sat above a plain of clouds but that was not the reason. After years of fear, John was sitting for the first time in an aeroplane.

Other passengers were more obviously excited – not exactly unruly but in exceptionally high spirits. They had come through against all the odds and were laughing with their fellow heroes. Triumph, infused with slight hysteria, prevailed. They slapped each other on the back, charged around kissing and shaking hands with near-total strangers, and one or two cried.

The crew encouraged the party atmosphere, urging passengers to take off their safety belts and walk around. Boisterous adults queued to see the flight deck. Cameras clicked. The aeroplane buzzed with laughter. There were cheers as the plane went through the clouds. Cheers for Mr Evans who had just celebrated his eighty-fifth birthday. Cheers for the pilot, cheers for passengers, cheers for any reason anyone could think of.

This extraordinary journey was solely for those afraid of

flying. The forty-five minute round trip from Manchester Airport was the culmination of a day's 'Fly with Confidence' course. This has been run by two British Airways pilots, Captains Douglas Ord and Peter Hughes, ever since 1986, when they realised that many on BA's flights to view Halley's Comet had no interest in astronomy. Nervous of flying, they simply wanted to try out a short flight. Since then, 10,000 have enrolled on the course and 98 per cent have boarded the aircraft at the day's end. No detailed follow-up has been carried out, but Hughes claims nineteen out of twenty feel more comfortable flying as a result.

Single-session treatments symbolise recent progress in dealing with phobias. Immersed in the latest therapies, sufferers can find their years-old phobia conquered in one day. Within a morning, they have been taken through the technicalities of flight, had their views on its dangers challenged and been taught basic relaxation techniques. In the afternoon this brief training is put to the test when they get on an aeroplane. Similar courses exist for those afraid of spiders. Various therapies prepare them for entry into the spider house. The vast majority are then happy to allow huge spiders to run up their arms and even through their hair.

These commercial courses represent the new attitude towards treatment for phobias. Directed at a few specific phobias, they are not a real option for most phobia sufferers. But they do demonstrate the prevailing optimism. In mainstream medicine, doctors and therapists are now confident that they can offer a working solution. Hundreds of years of theories and ideas have finally begun to make an impact.

A simple but dramatic shift in thinking has cleared the way. Phobias were once thought just the tip of an iceberg. Psychiatrists and psychologists believed that therapy had to be undertaken extremely gently for fear of what might be unleashed. This meant that treatment could involve months looking at words or drawings before moving on to the next stage. Vast patience and stamina were essential to complete such a course and most phobias continued unchecked.

In fact, though successful treatment can have knock-on effects, they are usually positive. Some people's self-esteem gets such a boost, for instance, when they overcome their fear of flying that it can improve all aspects of their life. One woman was completely distraught when she arrived. But fear of flying was only a fraction of her worries. She fretted about her journey home and what would happen if she was late to pick up her children. She was sure the ensuing chaos would make the family late to bed so they would all sleep in the next morning which would ruin the whole day. A catalogue of disaster stretched ahead. As it happened, she was so relaxed and delighted at her own success that she actually missed her motorway turn-off. Though late, the anticipated chain of events did not occur. Her new laid-back attitude persisted and she said later that the course had changed her life.

So phobias can be taken at face value and their symptoms addressed directly. This approach is standard in medicine. A girl with a bacterial infection normally improves quickly once on antibiotics. Her doctor does not have a philosophical debate about why she was the only child in the class to succumb. If she recovers quickly, the doctor has done a good job. Fear, like bacteria, can be confronted, and long-term inhibitions and preoccupations undermined almost immediately. This is tremendously encouraging for anyone with a phobia today. Treatments work and need not take for ever. The root cause may never be known but often has no bearing on the treatment.

Furthermore, phobias often disappear without trace. People suffering from depression may have to embark on a lifetime's struggle to keep symptoms at bay. Many ex-smokers or drinkers know that they remain a single lapse away from addiction. Phobias, in contrast, can be wiped out for ever with a single course of treatment.

Some phobias are replaced by a fascination with the thing once so dreaded. People previously afraid of snakes or spiders may keep them as pets. A fear of heights might be replaced by the new hobby of rock-climbing. Diana, whose story is told in chapter 5, once had such severe agoraphobia that she was unable to answer the telephone or open a letter. She now gives presentations about her experience. She will never forget her years with agoraphobia or how much it disrupted her (and her family's) lives, but she can no longer identify with the fear itself.

This new optimism should not, however, belittle the very real problems that remain unsolved. Phobias often go unheralded, unnoticed and, most importantly, untreated. The largest study ever, the Epidemiologic Catchment Area (ECA) Program in the US, aimed to discover how common various disorders are within the general population. More than 20,000 people in the community were interviewed about fifteen different disorders, among them phobias. Researchers found that fewer than one in four with a phobia had received treatment.

Treatments could be improved. Single-session treatments demand tremendous courage, too much for many. Thousands are helped but most phobics find the very idea of attending unthinkable. They often know about the courses, even acknowledge that they themselves could benefit, but their fear is far too great to allow them to sign up. People with phobias live with levels of unimaginable fear and this fear makes even a single day, which could eradicate their problem, too much to ask. Moreover, many who do attend are helped but not cured. Their terror is reduced but they remain exceptionally nervous. A rare few others are not helped at all.

Progress, though, is accelerating. Advances in psychology, psychiatry, genetics and molecular biology are all converging, improving our understanding of the causes of fear and providing new ways of addressing it. A range of much gentler approaches is now becoming available. Increasingly effective drugs can be used alongside behaviour and talking therapies. The talking therapies themselves are being streamlined to focus on practical improvements, rather than dwelling on childhood traumas. Improved understanding of how phobias develop may help us protect our children from ever developing them. Alternative, more speculative approaches suggest lifestyle changes which may make a difference. Whatever treatment is chosen, phobics can now expect relief from their fear in a limited number of sessions. This improvement has been one of the medical success stories of the decade.

## The Hidden Epidemic

So, is that the problem solved? Well, no, not quite. For a start, the scale of the problem is immense. Community studies report that up to two in five of us have a severe dislike of something, and even full-blown phobias are common. Strictly speaking, there is no phobia epidemic, since an epidemic usually refers to an infectious disease that has struck an unusual number. There is no evidence suggesting that there are suddenly more people affected than ever before. But to get a rough idea of the scope, let us just compare the prevalence of phobias with a relatively common infectious disease like influenza.

In the winter of 1999/2000, almost 400 out of every 100,000 in the British population went to their doctor with influenza, which means that it was approaching epidemic levels. The impact on the National Health Service was extreme. Hospital beds were filled with flu patients, so planned operations for cancer and heart disease had to be cancelled, and the NHS was in crisis. And this, at a level of 0.4 per cent.

Official figures are probably an underestimate, as many sensibly managed their illness at home, but even if only one in five consulted their doctor, that would mean the real figure affected was 2 per cent. How does this compare with the numbers affected by phobias?

The ECA study mentioned above found that between 4 and 11 per cent of interviewees had suffered from at least one phobia in the past month. From this, researchers estimated that more than 6 per cent of the population has a phobia at any one time.

We all knew at least a handful of people who were ill with influenza in the winter of 1999/2000. This guestimate suggests that, whether aware of it or not, we probably know three times as many with phobias. These figures give some idea of just how widespread phobias are.

## *A Few Famous Phobics...*

Phobias strike across the board, irrespective of intelligence, beauty or success. Kim Basinger and Sir Isaac Newton both suffered from agoraphobia. Sir Isaac was housebound for years up to 1684, after a period of severe stress. His mother died, a fire destroyed some important papers, he was exhausted after finishing his *Principia* and he was arguing with Cambridge University. All must have been distressing, but none restricted him as much as the agoraphobia that followed. Kim Basinger developed agoraphobia after the birth of her daughter. In some ways it is harder to imagine agoraphobia in an actress: academics can succeed with limited socialising, whereas actresses are subject to the most intense public scrutiny. But Basinger's experience of agoraphobia is typical of many women's. The hormonal and lifestyle changes surrounding childbirth are profound, and not even the most glamorous women are immune. This is explored further in chapter 8, on gender.

Arsenal striker Dennis Bergkamp has had a golden career. He has been voted FIFA's third best player and the top European. But he is unlikely to get a game in Greece, Turkey or Eastern Europe. For Bergkamp has a clause written into his contract ensuring that his club cannot insist on his flying. While the rest of the team take short flights to matches in the north of England or Europe, Bergkamp sets off by car, coach or train. He has flown in the past, but the last time was to play for Holland in the 1994 World Cup in the US. Since then, he has refused to fly at all and if he cannot get to a match overland, he cannot play. Bergkamp's fear is common knowledge in football circles but he will not talk about the reasons behind it. He has said that after he finishes playing football he may address his fear, but that for the time being at least, he is grounded.

Hans Christian Andersen was middle-aged by the time he developed his fear of fire, following the death of his old friend Jette Wulff in a blaze aboard the Atlantic steamer *Austria*. After that, Andersen always carried a rope with him, so that he could escape through a window in case of fire. He never used the rope, but it can still be seen at the Hans Christian Andersen Museum in Denmark. His behaviour was exceptional even at a time when fires were relatively common because contemporary buildings were often wooden. But his fear did not prevent him travelling, it simply added to his luggage. He wrote about fire in at least three stories, 'The Pixie and the Grocers', 'The Tin Soldier' and 'The Lovers', but he never tackled his fear.

He had other, stranger fears. He was afraid of dying, of seeming dead while still alive and of being buried alive. He was also afraid of seeing the dead. These fears were not unusual for the time. The mid nineteenth century was a morbid era and many were fixated with death. At a New Year's Eve party in 1845, he declared that dead people should mark their presence with tones. But then, both he and his hostess, Jenny Lind, the famous Swedish opera singer, were shocked and frightened when they heard a loud C ring out from an apparently untouched piano. However, he managed to capitalise on his fears. Twelve years later, in *To Be Or Not To Be* the hero hears a reverberating E and thinks it may be a sign from his dead beloved, Esther. So his fear was not entirely in vain.

## . . . And Some Fictional Ones

Spiders, snakes and rats are convenient symbols for fear or disgust and our screens are littered with them. Film directors rely on our near-universal unease to set a scene within a couple of frames. These are obviously difficult viewing if they are the object of your phobia and chapter 2 explores how far the media might even contribute to some of our fears.

Direct portrayals of phobias are less common but in the film *Arachnophobia*, Dr Ross Jennings (played by Jeff Daniels) has been intensely afraid of spiders all his life. His first memory is of himself lying near-naked in his cot when a spider crawled through the bars and on to his leg. His limbs froze and he was utterly helpless, unable to stop it moving over his bare skin.

The film climaxes with a replica of this incident. Now adult, Jennings is lying motionless, trapped by fallen rubble: the cellar is starting to catch fire. He watches horrified as a huge Venezuelan spider approaches his foot. This time the spider's bite would be fatal. It moves up his leg and onto his shirt. Is Jennings paralysed with fear again? Or does he have a plan? With impeccable timing, he waits until the spider climbs over a piece of wood lying across his chest, and then thumps the far end of the plank, catapulting the spider across the room into the fire. He and his country are spared.

George Orwell's *1984* tackles the subject of specific fears more directly. Winston Smith pales and endures 'a black instant of panic' when a rat appears in the secret room he shares with his girlfriend, Julia. Later, when his opposition to the all-pervading Party is discovered, he is sent to the Ministry of Love. He is beaten with fists, truncheons, steel rods and boots. He endures high-voltage electric shocks, is deprived of food and sleep, undergoes hours of questioning and makes numerous confessions, but still he loves Julia. Then he is transported to the notorious Room 101.

Room 101 contains 'the worst thing in the world'. His captor, O'Brien, tells him: 'The worst thing in the world varies from individual to individual. It may be burial alive, or death by fire, or by drowning, or by impalement, or fifty other deaths. There are cases where it is some quite trivial thing, not even fatal.'

This sounds rather like phobias. Winston's Room 101 contains two enormous hungry rats in a cage held close to his face. O'Brien continues:

> There are occasions when a human being will stand out against pain, even to the point of death. But for everyone there is something unendurable – something that cannot be contemplated. Courage and cowardice are not involved. If you are falling from a height it is not cowardly to clutch at a rope. If you have come up from deep water it is not cowardly to fill your lungs with air. It is merely an instinct which cannot be destroyed. It is the same with the rats. For you they are unendurable. They are a form of pressure that you cannot withstand, even if you wished to.

O'Brien is right. As the cage is brought so close that 'the foul musty odour of the brutes struck his nostrils' and the wire touches Winston's cheek, he starts shouting frantically: 'Do it to Julia! Do it to Julia! Not me! Julia! I don't care what you do to her. Tear her face off, strip her to the bones. Not me! Julia! Not me!'

With this betrayal, Winston's punishment comes to an abrupt halt, he is released but he is broken. He lives a humdrum existence until the 'long-hoped-for bullet' enters his brain.

The torturer's insights into extreme fear may be more telling than the routine assurances of health professionals. The idea of 'an instinct which cannot be destroyed' is untrue, as I will show, but describes how many feel when confronted with the object of their phobia. O'Brien also notes, correctly, that the cause of fear can be trivial and yet unendurable for that individual. He recognises that the fear is so intense that courage and cowardice become irrelevant and says that even people who could endure pain to the point of death will be unable to withstand it. Unlike health professionals, of course, he then goes on to exert just this level of pressure.

Alfred Hitchcock's *Vertigo* is another extreme phobic portrayal. Detective Johnny Ferguson (James Stewart) is chasing a criminal across wet rooftops. He slips and is dangling over an edge, clinging on by his fingertips. A colleague above leans down to him and offers a hand, but Ferguson is dizzy and unable to take it. The colleague loses his balance and falls to his death below.

This is Ferguson's first inkling of his vertigo and he quits his job with the police force. 'There's no losing it,' Ferguson is assured by his friend, Midge, who says that only another emotional shock will cure him. Rejecting this, he tries out some homespun behaviour therapy of his own, standing first on a stool and looking up and down. All is going well so he tries some higher steps. Unfortunately, he glimpses the street below out of the window and promptly faints. Thus his treatment ends. His vertigo is then assumed so permanent that others can base a murder plot on the certainty that he will not make it to the top of a tower at a crucial moment.

The three very different stories all successfully convey the extent of phobic fear and the individual cost. Ross Jennings, a highly respected doctor, has spent his life dreading spiders and relying on others to kill or remove them. Winston Smith betrays his girlfriend and, in the end, himself, through his fear of rats. Johnny Ferguson gives up a long-held ambition to become Chief of Police when he quits his job and, worse, is unable to save the life of the woman he loves because of his vertigo.

All three carry the fatalistic and depressing message that phobias are as much a part of us as our height or eye colour. It chimes with and may even have shaped the widespread perception that phobias are for life. Sadly, this is often true as we accept limitations on our lives far too readily and only a small proportion of phobias ever receive treatment. But, as this book sets out to show, phobias can be and are being cracked.

## From Antophobia to Zoophobia

So what do we develop phobias of? In short: anything. The National Phobics Society has a list of over 250 phobias, and it is not exhaustive. Some, like arachnophobia (fear of spiders) and claustrophobia (fear of enclosed spaces) are familiar; others less so. We may not know anyone with taphophobia (fear of being buried alive), antophobia (flowers), genuphobia (knees), metrophobia (poetry) or zoophobia (animals), but all these do exist.

Phobias are truly international, crossing the boundaries of language and culture. A study within the mainly Hispanic population of Puerto Rico relied on translated questions asked in the ECA. It found 12 per cent of people had phobias at some stage in their lives, a figure on the same range as mainland North America.

Studying phobias across cultures is more difficult: they were, for instance, once thought almost non-existent in sub-Saharan Africa. More recent work suggests that phobias are as common, just less obvious than in the West. Africans are likely to develop physical complaints as a result of fear and this can mask the underlying phobia. They also fear different things. Witchcraft, sorcery and supernatural phenomena are still important among peoples such as the Yoruba in Nigeria. Within Yoruba communities, people with no psychological problems routinely believe that others (who appear harmless) may be plotting against them. They cannot talk about specific concerns for fear of the sorcerer's retaliation. Nigerian research had to rely on drug-assisted interviews to break down some of this reluctance and found that at least 20 per cent of outpatients at psychiatric units were definitely phobic.

In the US, the ECA study estimated that between 1.5 and 12.5 per cent of the population has agoraphobia at some stage of their lives. Agoraphobia – literally, fear of the market place – usually translates to a fear of being away from home or a safe place. Using public transport, going to shopping centres or any crowded area is often out of the question. People with agoraphobia may become housebound, unable to work or have any sort of social life. Some are so anxious that they need someone with them constantly, even at home, which places a huge burden of responsibility on family and friends. The entire family set-up frequently revolves around the agoraphobia.

Agoraphobia is defined as a complex phobia because it is

often interlinked with generalised anxiety and fear. Most
sufferers are women. It typically starts after the late teenage
years and before the mid thirties, but can linger for years,
even decades. Arguably the most debilitating of all phobias,
it can touch every aspect of life.

Social phobia, another complex, all-pervading fear, was
found by Swedish researchers to affect between 2 and 20 per
cent at any one time, depending on the precise definition. It
is a fear of being scrutinised by other people and embarrassed.
Social situations, any sort of public performance, even eating
or drinking out, may be impossible. For some, anxiety is
limited to a single situation such as being unable to write in
front of others – tricky when most of us rely on credit cards
– being unable to speak in public or urinate in public toilets.
Well-defined social phobias like these may have an important
but relatively limited impact on someone's life. However,
like agoraphobia, social phobia can often have far-reaching
effects.

Social phobia is more evenly distributed between the sexes
and, if anything, more men than women are affected. It often
develops from childhood shyness, becoming full-blown in
adolescence, just as young people are starting to establish
their own social lives. Parties, eating out and shared activities
are a misery for those with social phobia. Dating can be a
nightmare. Solitary leisure pursuits and a career that avoids
any sort of public speaking are possible, but most families
and jobs demand some level of socialising. Some manage to
endure situations they dread, but their social anxiety effectively quashes all enjoyment.

This phobia takes varying forms in different cultures. In
Japan and Korea, people with social phobia do not worry
about being embarrassed, but are more likely to be excessively

afraid that they will offend others, either through body odour, blushing or eye contact.

Specific phobias, considered the least serious group of phobias, are more easily pinpointed and sufferers can say exactly what they are afraid of. Specific phobias often start in childhood and last a lifetime. They include fears of animals or insects, or of something in the natural environment such as storms, heights or water. Fear of blood, injections and injury come into this group, as do fears of specific situations such as tunnels, bridges or lifts. The same New York researchers estimated that one in ten of us has a specific phobia at some stage.

Specific phobias give flashes of extreme anxiety in set circumstances. They tend not to dominate lives as phobics may only need to avoid well-defined situations, such as lifts carrying more than six people beyond the tenth floor. But while phobias of buttons, wallpaper or cotton wool can sound trivial, bizarre or even funny, such fears can still affect career decisions or cast shadows over family life. A driver with arachnophobia could swerve dangerously if a spider appeared on his dashboard. Women with blood and injury phobias may decide not to have children because they cannot bear the thought of giving birth. Less dramatically, a fear of dogs can put a stop to picnics, and a fear of tunnels or bridges can make travelling extremely complicated.

Phobias fit into neat categories on paper, but in practice overlap and are difficult to distinguish. Someone who never goes out is probably agoraphobic, but may have social phobia if they avoid only social situations and fear being embarrassed in front of others. Someone terrified of buses or trains might have agoraphobia, but if they fear public transport and nothing else, it would be considered a specific phobia. Specific phobias exist of, say, dirty cutlery in restaurants, but someone who obsesses about dirt and has developed time-consuming cleaning rituals has an obsessive-compulsive disorder.

A single phobia such as fear of flying can have many roots and people on 'Fly with Confidence' courses tend to have mixed problems. Some are claustrophobic, some scared of heights; others are afraid of dying and convinced that flying is unsafe. The organisers estimate that a third of the attendees have never flown before but are terrified of the very idea. They are the easiest to treat, and some, like John, hardly need to get both feet inside the plane to feel better. Part of his fear was based on the assumption that planes are very cramped: one look was enough to disprove it. Another third have flown happily for years before having a bad experience which precipitated their phobia, either a physical event such as extreme turbulence, or a personal crisis which happened to coincide with a flight. The final third might still be flying regularly but feeling progressively worse about it. Their fear is likely to be part of a complex phobia – agoraphobia or social phobia – and they are the hardest to treat. They have seen inside planes, they know the statistics of aeroplane crashes, but no amount of information will help. Their fear is inside – they fear their own reactions, afraid of having a panic attack, terrified by the total lack of an escape route. They are sure that they will be the one running up and down the aisle, hammering at the door, screaming, 'Don't panic! Don't panic!' For these people, a quick-fix solution is unlikely to be all that is needed.

Clinical classification of phobias is important because of the seriousness of the complex phobias. Agoraphobia and social phobia routinely lead to missed opportunities in life but are also likely to be associated with other disorders.

People with social phobia are more than twice as likely to have problems with alcohol as non-phobics. Agoraphobics too are at an increased risk of alcoholism. Agoraphobia is also linked with some unfortunate personality measures such as dependency, unassertiveness and a lack of self-confidence as well as anxiety and avoidant behaviour. American research suggests that one in five with panic disorder, which is often associated with agoraphobia, attempt suicide. This is more even than people with major depression and twenty times the normal rate.

The tragedy is that phobias can be helped today but usually are not. Many different effective treatments exist but people continue to suffer. Phobias do have a stigma and it can be difficult to admit to a fear which you know, rationally, is out of proportion with reality. Why someone with a fear of heights should be more afraid of ridicule than someone with a broken arm or with high blood pressure is hard to say, though mental disorders traditionally have attracted less sympathy than physical ones.

Commercial one-day courses addressing fear of flying or spiders have the advantage of being based, respectively, at an airport or a zoo. Many find it easier to turn up there rather than at the local psychiatric unit and these courses have proved acceptable to those who might never seek help elsewhere. It seems likely, though, that their greatest appeal is among those whose phobias are least severe. In fact, drawing a line between normal fear and phobia is far from straightforward. Many, if pushed, would admit to disliking and fearing heights or spiders but do not have a phobia as this feeling causes no distress or infringement on their lives. A speaker at a recent meeting of the American Psychiatric Association pointed out that if she was not nervous about speaking to a room full of her most discerning peers, they might reasonably assume she was pathologically narcissistic. Some anxiety is not only inevitable in such a situation, it is probably good, prompting her to prepare her talk properly and deliver it well. If, however, she was so anxious about speaking that she refused to give lectures, or changed her job to avoid it, a diagnosis of social phobia would be appropriate.

## Up in the Clouds

On the 'Fly with Confidence' course, comradeship built up through the morning as people derived comfort from each other's questions and shared fear. The group started to bond. But as the day wore on they became more sub-dued, less friendly, some even angry at how ill-prepared they felt to climb aboard. People kept looking at their watches in alarm at how quickly the flight was approaching. During a desensitisation exercise, they were asked to imagine various scenes, such as checking in, waiting to board, climbing the steps to the plane. After each scene they were to return to deep relaxation. 'Impossible,' muttered the man on my right.

Despite such misgivings, all but one or two got on the plane and their relief was unmistakable. However, this was not universal. Across the aisle, a pale young man sat with his eyes closed, his head against the headrest. He was probably trying out newly learned relaxation exercises but he could have been praying. Helpful stewards provided numerous glasses of water, eliciting wan smiles, but did not make him much more comfortable. Occasionally he would open his eyes, look round, run his fingers through his hair and exchange a word with his neighbour. Then it was back to his private hell.

The man beside me seemed coolly confident, but confided that it was only OK because we were flying British Airways. The woman on the other side looked close to tears but chatted incessantly. 'I must be all right because I'm talking,' she said. 'If I was really bad I would be in a corner taking no notice of anyone.'

People are recovering in their thousands through courses like this, but the process is demanding. The pale young man only slowly regained his colour and back in the terminal he was still, inexplicably, clutching his untouched airline meal.

## CHAPTER 1

# History

## *In the Beginning*

The doctor was intrigued. His patient was in good physical health but he was so afraid of crowds and of the light that he hated leaving home. Whenever he went out, he chose if possible to go in the evening so that he could scuttle through deserted, dark streets to his destination. If he had to go out in the day-time, he would cover his head. He wanted to avoid seeing, and being seen by, anyone.

The man had done no wrong and had crossed nobody, but he was behaving like an escaped convict. He did not trust anyone outside his immediate circle. He was tremendously timid and the doctor became convinced that his fear of leaving home was due more to natural shyness than any real threat posed to him by the world at large.

The doctor was reminded of another patient who had yet another baffling fear. This man never went to parties, the theatre or any public gathering because he was convinced that he would disgrace himself. He thought he was bound to say something unacceptable, fall over or perhaps be sick in the middle of a crowd. Whatever it was, he believed that everyone would look at him, spit at him, jeer and mock him.

He was so sure that everyone hated him that he avoided public events at all costs.

The doctor mused over the two cases and went home and wrote in his journal about the 'men who feared that which need not be feared', a fair definition of phobias. The men's thoughts and behaviour will sound familiar today to anyone with experience of agoraphobia and social phobia even though the doctor was the Greek physician Hippocrates and he was writing 2,400 years ago.

Time has passed, language changed, but people's experience of phobias remains much the same. The first patient, according to Hippocrates, 'through bashfulness, suspicion and timorousness will not be seen abroad, loves darkness as life and cannot endure the light, or to sit in lightsome places, his hat over his eyes, he will neither see nor be seen by his good will.' The second, he said, 'dared not come in company for fear he should be misused, disgraced, over-shoot himself in gesture or speech, or be sick; he thinks every man observes him, aims at him, derides him, owes him malice.'

Hippocrates saw people with many different phobias over the years, ranging from agoraphobia and social phobia to animal phobias and other fears still common today. Damocles, he said, was terrified of heights and 'could not go near a precipice, or over a bridge, or beside even the shallowest ditch; and yet he could walk in the ditch itself'. He described other, quirkier phobias such as that of Nicanor, who was untroubled by the sound of someone playing a flute through the day but 'beset with terror' when he heard the same sound at an evening banquet.

Hippocrates' writing may be more poetic than modern medical notes but it demonstrates that the nature of fear has not changed over two thousand years. The ancient Greeks had the same experience of strange and unreasonable fears as we do today. Phobias have been around as long as we have, they are deeply ingrained in us, an integral part of human existence. This may not be much comfort to anyone with a phobia now but it does mean we have more than two thousand years' worth of thought and insight into fears and phobias.

Unfortunately, this does not mean two thousand years of steady advances in understanding. Periods of intense activity by scientists, philosophers and doctors have been separated by gaps of hundreds of years when little happened. Early insights were overtaken by other bogus or unhelpful ideas and progress has been as likely to move backwards as forwards. But sometimes an apparently new idea chimes with an ancient one. Many modern theories are updated versions of ancient thoughts and some of the questions that puzzled the ancient Greeks still go unanswered.

Hippocrates' careful observation of fear and phobias was exceptional at a time when most of his contemporaries thought that fear was sent down from the heavens. In Greek mythology, Phobos was the god of fright, son of Ares, the god of war. His brother was the god of fear, Deimos. Their companions included Eris, who represented strife and was insatiable in her fury; Enyo, who destroyed cities; and the Keres, who liked to drink the black blood of the dying. Myths related that this cheerful crew would stride on to the battlefield together, sowing disease and striking terror into the hearts of anyone they came across.

The god of nature, Pan, was responsible for contagious fear sweeping through crowds of people. Frightening sounds heard on mountains or in valleys at night-time were attributed to

Pan, and he was thought to be the cause of sudden, ground-less fear.

With the notable exception of the Stoics, the Greek people went along with mythology so far as to call on their gods for help and to blame them if they themselves were suffering. They would plead with Phobos to terrify their enemies, and at the same time assume that he was causing their own fear. They thought that Pan could determine the outcome of wars by generating mass hysteria throughout the ranks of one or other side and causing whole armies to disintegrate.

The Greeks were clearly comfortable with the concept of different types of fear. Phobos represented a sudden and acute fright, different from Deimos' ongoing, rumbling fear. Pan symbolised the sort of fear which can spread through groups of people. This classification has been modified over thousands of years but still exists, another clue that our experience of fear has not changed much.

The words we use to describe these emotions reflect the ancient beliefs. Today's Greek word *phobos* means intense fear or terror and translates directly into our word, phobia. The word panic is derived from Pan and has shifted its meaning more recently. It was once used to refer to the group process of mass panic, but now refers to an individual's experience, including panic attack or panic disorder.

Our word anxiety comes from the Latin *anxietas*, which means troubled in mind. Again, the meaning has held steady despite translation into numerous other languages. French, Italian and Spanish all contain words derived from the Latin. *Anxo* in Greek means to squeeze, embrace or throttle, which came to mean weighted down with grief, burdens and trouble and has passed into German as *angst*. The sensation of constriction or tightening across the chest, of being unable to breathe freely, is a classic feature of panic.

Agoraphobia may have been described by Hippocrates, but it was not given the name until much later. The German psychiatrist C. Westphal coined the term '*die agorophobie*', in a paper published in 1871. He described three men who either could not walk alone through certain streets or squares, or could do so only with great anxiety or a couple of stiff drinks inside them. Thinking about the feared situation could be every bit as alarming as actually walking into it. Westphal wrote:

> The patients derived great comfort from the companionship of men or even an inanimate object such as a vehicle or cane. The use of beer or wine also allowed the patient to pass through the feared locality with comparative comfort. One man even sought, without immoral motives, the companionship of a prostitute as far as his own door.

Westphal's choice of name harks back to Hippocrates' time when the *agora* was a public meeting place, used for discussions of public affairs, games or contests. In ancient Greece a contest could be athletic, poetic or a mental challenge between dramatists, and was known as an *agonia*. An *agonia* demanded that individuals tested their skills and later the word came to mean mental anguish. After Westphal, confusion arose and agoraphobia came to mean not only a fear of open spaces or public places, the *agora*, but also the fear of deficiency in one's performance, or *agonia*. It was not until the 1970s that the term 'social phobia' was brought into use to refer distinctly to the second of these, the fear of public scrutiny. It took a long time to get back to the starting point. Hippocrates may not have named these fears, but he certainly described the difference between agoraphobia and social phobia.

Hippocrates also explored possible causes. Unlike most Greeks, he thought it ridiculous to blame the gods for fear. He insisted that there was a physical cause within the individual. Neurotic symptoms fell into the class of melancholia, a type of insanity. It was caused by a build-up of black bile which made the brain overheat and caused passing terrors. Treatment was a regimen of diet, activity and exercise, designed to rid the body of the excess black bile. If this was not successful, drugs such as the poisonous hellebore were often given. The resulting vomiting and diarrhoea were taken as signs that the bile was being eliminated.

Hippocrates' confidence in this particular scheme was somewhat misplaced but his belief in a physical cause for mental disorders has been shared by scientists ever since. One of his younger contemporaries, the philosopher Aristotle, also searched for a physical cause for nervousness. Aristotle decided that the heart was the seat of all sensations and the brain a cold, bloodless part of the body which absorbed hot vapours arising from the heart. This led much later to the old English idea of 'the vapours', meaning a nervous disorder, low spirits or boredom.

Great Greek thinkers and twentieth-century neuroscientists may be united in their belief in a physical, biological cause of fear but there have always been other ideas. The Stoic school of philosophy grew up shortly after the time of Hippocrates and survived for five hundred years, well into the Roman empire. The Stoics included emperors (Marcus Aurelius), slaves (Epictetus) and even Nero's tutor, Seneca.

Stoicism stressed the importance of human reason in finding an accord with nature. Emotions had to be conquered and passions shed in order to achieve imperturbability. People can be happy in the midst of the severest pain if they can master themselves and let nothing overwhelm them. We are not at the mercy of external events. (Cognitive therapy (chapter 6) still relies on some of these ideas.) More specifically, the Roman Caelius Aurelianus wrote that phobias were a type of mania and arose from problems in the mind, not from the body or physical brain:

Mania fills the mind now with anger, now with gaiety, now sadness, now with nullity, now with the dread of petty things. As some people have told; so that they are afraid of caves at one time, and chasms at another, lest they fall into them; or there may be other things which frighten them.

More than two thousand years ago, then, philosophers and medics could give a good description of phobias. Ideas about the causes may have been primitive but they were forerunners of some of the main schools of thought still in existence. Sadly, the brilliance of these great thinkers probably had little impact on most people of the day. The prevailing view was that fear was sent down from the heavens and that phobias were best treated by trying to appease some god.

The Roman empire, which had assimilated Greek civilization, itself collapsed in about AD 400. The Church then dominated society and effectively put a halt to studies into individuals' emotional experience. Phobias obviously still existed, and fears of plague or syphilis were especially common. However, in a backward step for science, excessive or strange fears were assumed to be caused by an interaction between forces of good and evil, and people with phobias thought to have been overtaken by demons or evil spirits.

Throughout the Middle Ages, the Church dominated scholastic thought and pre-eminent thinkers were occupied with big theological questions. Not until the fourteenth century did attention turn gradually back to the individual. This paved the way for the golden age of philosophy, out of which grew psychology as we know it today. And it started with Descartes, once described as the first modern man.

## Cartesian Logic

Born into a rich and noble family at the end of the sixteenth century, Descartes studied languages, literature and philosophy at one of the top French schools of the period. But even as a young man he became disillusioned with the limited nature of the teaching and quit his studies to lead a life of pleasure in Paris.

Boredom eventually set in and he joined the army in Holland, where he learned about mathematics and the natural sciences. Then later on, he joined the Bavarian imperial army in the Thirty Years War, which allowed him to travel through Germany, Austria, Hungary, Switzerland and Italy. He was constantly observing, contemplating reality, and working on his own philosophical method.

He moved back to Paris for a few years, eventually leaving again for a life of near-seclusion in Holland. His interests included mathematics, optics, astronomy, chemistry and botany and out of this unlikely mix came key ideas in the history of psychology.

His views on early learning, for example, are still vividly contemporary. In 1649, shortly before his death, he wrote that learning can start before birth.

It is easy to conceive that the strange aversion of some, who cannot endure the smell of roses, the sight of a cat, or the like, come only from hence, that when they were but newly alive they were displeased with some such objects, or else had a fellow-feeling of their mother's resentment who was so distasted when she was with child; for it is certain there is an affinity between the motions of the mother and the child in her womb, so that whatsoever is displeasing to one offends the other; and the smell of roses may have caused some great headache in the child when it was in the cradle; or a cat may have affrighted it and none took notice of it, nor the child so much as remembered it; though the idea of that aversion he then had to roses or a cat remain imprinted in his brain to his life's end.

Descartes' major contribution applies to the whole of psychiatry, not just to phobias and anxiety. However, the respect in which he was held rather unfortunately cast in stone the mind–body split still so relevant to the treatment of phobias.

Descartes set out to question all accepted wisdom and build up his own philosophy from scratch. He was a firm believer in reason and thought all experience was fallible for he could never be completely sure that he had not been dreaming, or even tricked by a malicious demon. Bodily experiences were unreliable, he said, and the only thing he could be absolutely sure of was that he was thinking. His first principle of philosophy was, famously, '*Cogito, ergo sum*', 'I think, therefore I am', and he came to regard the mind or soul as totally separate from the body. They are simply two different entities, he said: the mind is not a physical thing and therefore it can never truly merge with the body.

Descartes was searching for absolute truths and was not attempting to pit future psychiatrist against future psychiatrist. However, his reasoning led to Cartesian dualism, which has translated into medical circles as the great divide between mind and body. Does the cause of a psychological problem such as a phobia lie in the thinking mind or in the physical brain? The question has never been answered and professionals line up on opposite sides of this divide. Geneticists, molecular biologists and neurophysiologists, looking ever more closely into the physical and mechanistic workings of the brain, represent the 'body' side of the argument. Their remit is to explore the parts of the brain that can make us susceptible to phobias, anxiety and panic, somehow change its delicate chemistry and reduce our fear. On the 'mind' side, psychologists and psychotherapists examine past experiences or current beliefs and aim to challenge and change our thinking patterns to dispel our phobias.

Descartes believed that mind and body were closely linked and he would not have supported this interpretation of his work. In *The Passions of the Soul* he wrote: 'There is such a tie between our soul and body that when we once have joined any corporal action with any thought, one of them never presents itself without the other.' It is ironic that a philosopher who gave himself the widest possible brief is best remembered for naming the rift between some of the most polarised views in medicine and psychiatry.

He even named the meeting place between mind and body

as the pineal gland. We now know that the pineal is sensitive to light and one of the hormones produced there, melatonin, regulates our sleep–wake cycle. Scientists researching jet lag and shift-work patterns have long been interested in the pineal but their work apparently had little relevance to phobias. However, some researchers now believe that certain light frequencies, acting via the pineal, may influence our susceptibility to both anxiety and phobias (discussed further in chapter 9).

Descartes' belief in the central nature of thinking and reason makes him, like the Stoics before him, a rationalist. Cognitive therapists say that our beliefs fuel our fear, almost 'I think, therefore I am frightened.' Chapter 6, on cognition, examines at this in depth and it is quite possible that Descartes would have agreed with some of the main ideas.

Immanuel Kant, more than a hundred years after Descartes, was another rationalist, and his ideas fit equally well with cognitive therapy. Again, he stressed the importance of reason. He said, 'The understanding cannot see. The senses cannot think. Only by their union can knowledge be produced.'

Kant believed that our ideas shape our view of the world. It is as if our ideas are spectacles that distort what we see. They determine what we focus on and how appealing it looks. We do not see an event itself, but only its appearance through these unreliable glasses. Put simply, there are alternative ways of looking at any event. Cognitive therapists today would agree. They aim to change people's interpretation of events, just like adjusting their spectacles to change the focus or the tint.

## Locke and Empiricism

An alternative view is that reason does not come into it at all. The human mind is, in fact, like a blank sheet of paper. Ideas are generated through our physical senses and our experiences, and projected on to this blank sheet. We work on the information derived from our senses, make associations and generalisations and build up our psychological picture of the world. No matter how abstract or complex the idea, it begins with physical sensations. Even belief in the existence of God can be built up in this way.

These are the thoughts of John Locke, who was working soon after Descartes. He belonged to the opposite tradition in philosophy, empiricism, which rated experience above all else.

Learning and memory are built on experience alone, Locke said. Phobias are therefore learnt as the result of a bad experience. And like behavioural therapists today, he said that fear can also be unlearnt through experience. Locke's most important work was the *Essay Concerning Human Understanding*, published in 1690. His advice on dealing with irrational fears could have come straight out of a modern behaviour therapy textbook:

If your child shrieks and runs away at the sight of a frog, let another catch it and lay it down at a good distance from him; at first accustom him to look upon it; when he can do that, to come nearer to it and see it leap without emotion; then to touch it lightly, when it is held fast in another's hand; and so on, until he can come to handle it as confidently as a butterfly or sparrow.

Empiricists like Locke would be at home talking to behaviour therapists in the twenty-first century.

Behaviourism aligns itself with empiricism in the same way that cognitivism is linked with rationalism. These two schools of thought have continued through history like parallel lines, never getting any closer to each other. They ask different questions and look for different answers. Empiricists and behaviourists want to know *what* a boy does if you send a dog into the room where he is sitting. Rationalists and cognitivists ask *why* the boy starts screaming and climbs out of the window.

Unfortunately, accurate observation and brilliant insights did not necessarily translate into practical and humane treatments. The seventeenth and early eighteenth centuries may have been a time of rapid progress in theories of learning and thinking, but people with anxiety disorders probably shrank from some of the proffered cures. The physician Thomas Sydenham suggested that hysterical disorders or today's anxiety disorders could be helped with 'bleeding, purging, opiates, foetid medicines, chalybeate medicines, filings of steel and rhenish wines, plaister at the navel, hysteric julap, opening pills or electuary'. Another writer regarded anxiety as a symptom of cardiac disease and recommended 'narcotics and anodynes, mucilages, things fat and emplastik, emulsions and roborants'. Royalty escaped lightly in comparison, and Queen Anne's physician Sir Richard Blackmore used opium because, he wrote in 1725, 'It calms and soothes the disorders and perturbations of the animal spirits.'

Treatment provided by the clergy was, perhaps predictably, more orientated towards 'mind' than 'body'. One minister thought that the key to treating anxiety was to 'put them in a Pleased condition'. Another clergyman who specialised in

'Spiritual physicke to cure the diseases of the soule' was effectively an early psychotherapist and recommended the use of silence.

## Darwin and the Dawn of Modern Science

Charles Darwin had no desire for a head-on collision with the Church. By nature he was diplomatic and unassuming, certainly not confrontational. On top of this, his beloved wife Emma Wedgewood was deeply religious. But he was exasperated by the Church's stranglehold on biology. The doctrine that we are made in the image of God, implying our perfection, was a particular problem. It elevated any study of humans into a direct challenge to God's greatness and effectively stifled scientific thought.

Scientists in other disciplines in the early nineteenth century had much more freedom than biologists. Darwin viewed them jealously as he wrote: 'What would the Astronomer say to the doctrine that the planets moved [not] according to the laws of gravitation, but from the Creator having willed each separate planet to move in its particular orbit?'

Another problem facing biology was the legacy of rationalist philosophy. It had produced great insights and set up trains of thought still followed today, but in practical terms it had come to a dead end. The philosophical view of the brain did not lend itself to systematic study. Kant even said that the mind was unquantifiable and beyond direct investigation so that a science of the mind was a logical impossibility.

Darwin eventually overcame these obstacles and his work

paved the way for an explosion of activity in mind and brain research. The parts relevant to phobias are explored in detail in the next chapter, but his greatest contribution was, in the mildest possible manner, to wrest control of biology from the Church.

*The Origin of Species*, published in 1859, simply observed that living things adapt themselves to their surroundings. Species change over generations, he said. And if living things are not God-given, created once for all time, this implies that they are a legitimate target for scientific study. Darwin carefully excluded humans so as not to court more problems with the Church than was absolutely necessary, but his argument plainly implied that we are not distinct from animals. And if we have not been selected by God for special treatment, there is no reason why we cannot be studied scientifically.

The work caused a social and moral storm on publication but it was assimilated by scientists and the public alike within a decade. From there, developments in the many fields of science relevant to anxiety, fear and phobias started to snowball. Psychologists started collecting data. They devised experiments and studied the behaviour of animals and humans. Wilhelm Wundt, who set up the first psychology laboratory in Leipzig, epitomised this new approach. His followers were trained to look for traits that could be measured, and then collect data before they started building their theories.

The foundations were laid for the modern neurosciences. Spanish physiologist Santiago Ramón y Cagal was awarded the Nobel prize for his discovery that the brain is made up of neurones, or nerve cells. Afterwards, scientists went on to establish that the brain consists of a vast interconnected network of these cells. Scientists today are still trying to determine how far communication between these cells determines emotions like fear or anxiety.

Medicine progressed. Psychiatrists like C. Westphal started closely observing and defining phobias. The American Civil War brought tragic opportunities for clinicians to study fear at close quarters. Doctors who might once have moralised about courage or faith started taking measurements and looking more dispassionately at the effect of fear on the heart, lungs and other body systems. They wrote up their observations and developed theories, some of which are explored in chapter 4, on neurophysiology.

All of this work set the scene for probably the single most influential figure in the history of thought on phobias.

## Freud on Fear

Five-year-old Hans was walking along the street with his mother. It was in the early years of the twentieth century and there were horses trotting by, pulling carriages, vans and buses. Wheels and hooves clattered on the cobbled road, people walked across and between them and it was an ordinary, busy day.

Hans' mother stopped to greet an acquaintance and he looked idly down the street. A bus came towards them, pulled by two black horses. Just as it reached them, one of the horses stumbled and fell heavily on to its belly, jolting the bus. The other horse swished its tail in distress, shook its head and stamped its feet. The driver jumped down to help, and people peered out of the bus to see what was going on.

The horse looked enormous as it lay on the ground,

covered in dust, in front of Hans. It snorted and tried unsuccessfully to get up. Its harness was twisted, making it difficult for the horse to move. A crowd started to gather and the noise grew. Hans stared in fright and his mother led him away from the scene.

Hans was agitated and could think and speak of nothing else for the rest of the day. Soon afterwards, he developed a phobia of horses. He started crying on his normal walk to the park with his nursemaid, and wanted to be taken home. He was afraid a horse would bite him and later, at home, afraid that a horse would come into the room. He became reluctant to go out at all and was quiet and withdrawn in the evenings.

Hans led an unexceptional, if somewhat closeted, middle-class life. He had a younger sister but few friends of his own age. His parents had, for the previous two years, noticed what they considered to be a precocious interest in sex. He asked questions about his parents' and his younger sister's genitalia. Both parents told him that he had the phobia because he often touched his 'widdler'.

His parents were devotees of Sigmund Freud and Hans's father wrote to Freud about the horse phobia. The case of Little Hans, as it became known, was the first published account of child analysis, and became celebrated as a key success in psychoanalysis. Freud believed it was proof that sexual urges are an essential part of the development of phobias.

Freud said Hans had an Oedipus complex. He loved his mother and wanted to be taken into her bed. He wanted his father, his rival, dead, and was therefore afraid of him. Hans transferred his fear of his father on to horses and the phobia erupted after some months of longing for sexual contact with his mother. According to Freud, 'His sexual excitement suddenly changed into anxiety.' The affection for his mother 'triumphantly achieved its aim', by making Hans afraid of going out into the street and allowing him to stay at home with his mother.

These were tendencies in Hans which had already been suppressed [wrote Freud] and which, so far as we can tell, had never been able to find uninhibited expression: hostile and jealous feelings against his father, and sadistic impulses (premonitions, as it were, of copulation) towards his mother. These early suppressions may perhaps have gone to form the predisposition for his subsequent illness. These aggressive propensities of Hans found no outlet, and as soon as there came a time of privation and of intensified sexual excitement, they tried to break their way out with reinforced strength. It was then that the battle which we call his phobia burst out.

Hans's father received Freud's letters over a period of months and Hans recovered gradually.

The case of Little Hans, which has been argued about ever since, was a fascinating example of Freud's theory of mind. Freud said the 'id' stands for untamed passion, and the 'ego' for reason and good sense. The id generates inappropriate sexual or aggressive impulses, which the ego tries to make socially acceptable. The ego receives the id's dangerous urges, and represses them if at all possible. Failing that, the ego has to find another outlet and may alter excessive excitement so far that it ceases to be pleasurable and becomes a psychosomatic symptom such as palpitations or fainting.

So, according to Freud, Hans's sexual desire for his mother

and his longing for his father to be dead were unacceptable urges created by the id. They created more excitement than the ego could deal with and were converted into anxiety and a fear of horses.

A single anxiety attack, with palpitations or breathlessness, can be enough to trigger a phobia, said Freud, and many today would agree with this part of the theory, if not with the underlying explanation. The ego feels helpless during the attack, Freud says, and deals with this by projecting the anxiety on to the external situation. The anxiety attack then becomes associated with a bus or train journey, for example, rather than with forbidden excitement. It is easier to control external situations than internal thoughts and the phobia becomes a convenient way of explaining away a terrifying reaction. This process is carried out at a subconscious level; at the conscious level, all we know is that we are afraid of the situation. The ego then prompts us to avoid the situation in future as a defence against the impulse and the feeling of helplessness. So an external, tangible fear replaces an internal, instinctual danger and a phobia develops.

Freud's theory was at once brilliant – and mistaken. On the positive side, he founded psychoanalysis and highlighted, long before mainstream science, the importance of relationships and childhood experience in shaping adult personalities. Furthermore, his emphasis on the subconscious opened up a whole new way of thinking about mental pain and inner conflicts. But his emphasis on sex was questionable. Other scientists have combed the 140 pages' worth of his analysis of Little Hans and found no evidence that the boy sexually desired his mother. Nor are there any signs that he feared his father. But the link between sex and phobias, started by Freud, has hampered phobia treatment for decades.

Specialists and theorists today have largely given up the idea but it still lingers and can cause problems. Referral for a phobia from GP to psychiatrist can still lead to a discussion about relationships with parents and possible abuse. It is impossible to completely rule out a link between sex and a phobia; complex phobias can develop from a background of troubled emotions and difficult experiences and there are, of course, phobias of sex itself. But most phobias do not stem from sex or have anything to do with it.

Psychoanalysis itself is largely out of fashion in scientific circles, partly because it has failed to keep up to date and incorporate new findings from cognitive science or physiology. But its legacy in phobia treatment is the enduring idea that phobias are the visible signs of terrible inner conflict. For many years psychiatrists have approached phobias with extreme caution for fear of what might be uncovered.

The whole basis of recent progress (and this book's premise), is that delving in someone's unconscious, looking for the cause of a phobia, does not get results. Good treatments work regardless of the cause.

Other, less celebrated parts of Freud's work provided a solid framework within psychiatry which still largely exists. He had a genius for making sense of the experiences described by his clients and categorising the different aspects of anxiety. The remnants of his classification are still apparent today in the mighty *DSM*, the *Diagnostic and Statistical Manual of Mental Disorders*, the bible of the American Psychiatric Association.

Until Freud's paper in 1894, anxiety disorders were collectively known as neurasthenia. Freud's achievement was to distinguish 'anxiety neurosis', or what we now call panic disorder, from general anxiety and to describe its three aspects. The first is the anxiety or panic attack; the second is anxious expectation, or anticipatory anxiety; and the third is phobic avoidance. Further, Freud recognised that people could have more than one anxiety syndrome, and that people could have mild forms of anxiety.

He described heart 'spasms' and the difficulty in breathing that can accompany anxiety attacks. In fact, ten of the thirteen symptoms included in the *DSM* as late as 1987 had been previously noted by Freud. Freud also recognised that specific phobias – not the term he used – were quite different from agoraphobia, and described them as an exaggerated reaction to dangers instinctively feared by everyone. The idea that most specific phobias are an overblown but essentially normal reaction surfaced again recently and is explored in the following chapter.

Freud, who was a neurologist, drew together much of the work that preceded him and his early work reflects his interest in neurophysiology. He cut across the nature–nurture debate and claimed, in a truly modern fashion, that inborn and biological factors interact with experience in causing anxiety disorders. Interestingly enough, he was himself phobic about travel.

## *Little Albert*

J. B. Watson was an impatient young American. He was irritated by the state of psychology in the early twentieth century, by its lingering obsession with philosophical questions and its fascination with the subconscious. He set out to drag it into the realms of science.

Watson did not claim that mental phenomena do not exist,

but rather that they cannot be measured and therefore might as well be ignored. In an unlikely agreement with Kant, he said that the mind, or consciousness, could not be investigated scientifically. Following on from Wundt, mentioned earlier, Watson stressed the importance of collecting data and measuring overt, visible behaviour. Little Albert, an 11-month-old baby, was the unfortunate guinea pig chosen.

Watson's masterstroke was a direct challenge to Freud. He and his colleague, Rosalie Rayner, allowed Albert to play happily for a while and then showed him a furry white rat, at the same time banging an iron bar on metal just behind his head. The little boy got a terrible fright. A few days later, they showed Albert the rat again, this time without the noise. He was still obviously frightened. In fact, weeks after the experiment, he remained afraid of rats, dogs and anything furry, even fur coats.

A single, frightening event was enough to create a lasting fear in Little Albert. By extension, it suggests that the horse's fall in front of Little Hans may have been sufficient in itself to cause his subsequent phobia. Analysis of his subconscious was therefore unnecessary.

Russian neurophysiologist Ivan Pavlov, working at the same time, would have agreed. He famously rang a bell every time he fed a group of dogs. Eventually, the dogs started to salivate at the sound of the bell whether or not there was any food. Pavlov said they had come to associate the bell with the food so strongly that either would make them salivate. The dogs were conditioned, to use Pavlov's term, to salivate when the bell rang.

White rats were a convenient vehicle for studying behaviour because, like dogs, they can be conditioned. Simple experiments with rats produced simple results and fuelled enthusiasm for behaviourism. In variants of Pavlov's experiments, rats were shown something innocuous, like a coloured light, at the same as they received a mild electric shock. With repetition, the rats came to fear the light alone.

It provided a simple way of thinking about phobias. A single event causes lasting fear. A child is frightened when a big dog snarls and attempts to bite and afterwards fears and avoids all dogs, even small and friendly ones. Behavioural therapy attempts to reverse the process. By gradually reintroducing the child to dogs, the link between the snarling dog and others is broken, the child gains confidence and the fear disappears.

However, behaviourism failed to see off Freud. Its practical shortcomings were, ironically, demonstrated by Watson. Like Freud, he was unable to heal himself. Watson had a lifelong fear of the dark which his behaviourist methods could not banish. It is hard to imagine anyone with a phobia believing as fervently in their treatment or being as determined for it to succeed, but it did not work for Watson.

His personal life may have dealt behaviourism an equally serious blow. He had a scandalous divorce following his affair with Rosalie Rayner. They subsequently married, but he was forced out of his job at the prestigious Johns Hopkins University and left academia for advertising. Behaviourism was robbed of its figurehead, research started to go in many different directions and it never regained its earlier theoretical coherence.

Behavioural learning theory may have foundered but behaviour therapy, a logical extension of the theory, is still a core feature of most treatments for phobias. Just as Watson was only interested in studying behaviour, the task of the modern behaviour therapist is limited to changing behaviour.

Watson did not argue that consciousness did not exist, only that it could not be measured. Similarly, therapists acknowledge that phobias mean fear, but they do not tackle the emotion directly. Instead, they work to change behaviour and prevent avoidance of the feared object. The therapy, discussed in chapter 5, may not have helped Watson, but it is routinely successful.

## Computers, Cognitivism and Progress

Freud and Watson's pre-eminent positions were eventually usurped in the 1950s by the silicon chip. Computers provided the inspiration for the next way of thinking about thinking. Centuries earlier, doctors trying to understand the heart were baffled until engineers invented the pump. The pump gave them a model for how the heart could work, and it was a good comparison. In the same way, computers introduced notions of information processing and storage which were new. Doctors hijacked these ideas to explain the workings of the human mind and memory. The analogy of programming a computer to carry out tasks was a more satisfactory explanation for how we learn complex skills than anything behaviourism had put forward.

So computers' first contribution to the progression of thought on phobias was as a model for thought processes and the mind. More recently, computers have driven research into the physical causes of fear in a way that has never been possible before. The power of modern computers allows geneticists to trawl through immense heaps of data in an attempt to pinpoint the genes responsible for panic disorder.

Advances in imaging have given scientists new ways of looking at the brain and allowing them to piece together an
ever-clearer picture of the physical changes when someone
thinks, laughs or is afraid. Computers are being used to
design molecules that will surely give us the next generation
of fear-busting pills. The neurosciences are advancing in
many different directions and none of it would be possible
without the modern computer.

But if history has taught us anything, it must be that no
one school of thought has all the answers. Hippocrates and
the ancient Greeks gave telling descriptions of phobias, but
did little to help their patients. The philosophers had some
brilliant insights which failed to make it into clinicians' text-
books. Progress has been halting over the centuries, charac-
terised by dead ends, false dawns and the odd piece of
brilliance, quickly obscured.

Freud must be credited with separating phobias from generalised anxiety and establishing them as a new and distinct
subject for study. His classification of anxiety types was a big
step forward because it drew attention to the special and
specific features of phobias which today's therapists are capi-
talising on and which are making new treatments so promising. However, his later psychoanalytical work, with its
subjective interpretations of hidden feelings, is spectacularly
unfashionable in the age of computers and hard data.

Practical progress has been most marked where sciences
have interacted. The theory of cognitivism may have overtaken behaviourism, but most phobia clinics now offer
cognitive–behavioural therapy. The theories may be irrecon-
cilable, but the two approaches taken together are more
effective than either alone.

This could also be true of the neurosciences, psychology

and psychoanalysis, which continue to pay scant regard to each other's findings. They have developed more or less independently, with little reference to each other. Neuroscience is a thriving field at the beginning of this new century and it is tempting to feel we can safely reject everything that has gone before. Undue attention to underlying problems, spearheaded by Freud, held back treatment of phobias for years and the demise of psychodynamics has been liberating and productive for scientists, clinicians and those with phobias. It is thrilling to be able to discard confusing psychoanalytical theory in favour of neuroscience and its promises of definite answers to clearly defined problems. But the dawn of the neurosciences could yet produce a need for a deeper understanding of the meanings of fear. Because, in the end, fear is more than a chemical reaction. No one with a phobia really cares about their hormone levels or brain activity. What they want is an end to their phobia and the sensation of fear.

## CHAPTER 2

# Evolution

## *Living Without Fear*

The man stood, arms outstretched, looking at the traffic below. He grinned, threw back his head and laughed. The wind ruffled his hair and tugged at his coat and he seemed euphoric. He started to turn clumsily round and round on the spot, like a small child having fun. A few yards away, his wife stared at him in disbelief. He was dancing on the corner of a parapet on the roof of a San Francisco skyscraper, one step from certain death.

The fictional character Max Klein, played by Jeff Bridges in the film *Fearless*, had survived a plane crash and became convinced he was invulnerable. His high-rise jig came some time after he walked across a city highway without looking, cars and vans screeching to a halt all around. He ate a bowl of strawberries, knowing that his allergy to them could cause a fatal reaction. Finally, he drove at top speed into a brick wall.

Klein survived a few months of this behaviour, but his life was disintegrating. His close encounter with death during the crash had eliminated his day-to-day anxieties and he felt he did not have to answer to anyone. He became so self-sufficient, not to say arrogant, that he felt little need for the closeness of those around him. He was remote and distant from his wife; he alienated friends with his lack of sensitivity. He spent more time with a young boy he had rescued from the crash than with his own son. He was not working, but spent his days looking at buildings. When introduced to a fellow survivor of the crash, he told his wife he had a feeling of overwhelming love for this woman. He had never felt anything like it before, he said. A few months of this and his wife was ready to leave him.

His psychiatrist was struggling with an extreme case of post-traumatic stress disorder; Klein himself claimed the crash was the best thing that had ever happened to him. It had been extraordinary, and had shown him 'the taste and touch and beauty of life'. He would not give up this state of mind.

Subjectively, Klein felt more alive than ever; objectively, he stood to lose his wife, son and home, his friends and his livelihood. It is an interesting take on fear. We are so used to portrayals of neurotics crippled by a million anxieties that we seldom stop to think what would happen if we had none at all. Anxiety and stress have a bad image. They are the scourge of the modern age, blamed for everything from undermining happy marriages to destroying sleep and causing headaches. Anxiety exaggerates bodily pains, it ruins good performances at work or school and quenches joy and laughter. It leads to alcoholism, eating disorders, domestic violence. The lifestyle pages of newspapers and magazines are filled with articles about dealing with stress and, we are told, life without anxiety would be wonderful.

Yet in this film, fear is portrayed as the glue that holds lives together, keeps marriages, friendships and careers intact and protects us from avoidable accidents. Klein eventually realises he needs help, regains normal sensitivities – along with his allergy to strawberries – and the story is resolved.

Anyone behaving like this in real life would be lucky to escape so lightly.

The anxiety system can go wrong, of course, and we would all like to banish the misery of panic attacks, obsessional behaviour or phobias. Successful treatment for these problems can revolutionise lives and nobody wants to get in the way of this. But evolutionists insist we would benefit from taking a step back and looking at why there is so much anxiety in society. They challenge the prevailing view of anxiety as a wholly negative experience. On the contrary, they say anxiety is a prime motivator, a positive drive, a force for good. It prompts us to achieve at work, to guard our reputation and to keep our families together.

We do not doubt that other animals need the ability to recognise and respond to threats. All living things face danger and must react appropriately if they are to survive. Creatures have a fascinating array of defence mechanisms, each specific to the threats they most commonly encounter. The chameleon changes colour to blend in with its surroundings and hide from potential attackers. A threatened squid squirts ink at its aggressors. Antelope simply run away from lions. Moths are preyed on by bats and have become experts in bat-frequency signals. They monitor the signals continuously and map the direction of their predators' flight. Only if the bat is heading directly for it does the moth snap its wings shut and fall, as if dead, to the ground. Familiar reactions like these have been sufficient, not for every animal to survive, but to keep the species going.

The intensity of the reaction also has to be appropriate since animals use up precious resources when trying to defend themselves. An antelope that is too ready to give up grazing and run will soon become undernourished; squid do not have unlimited ink. Even the simplest creatures have remarkably sophisticated responses, as demonstrated by American biologist Herbert Jennings, working in Europe at the turn of the century.

Jennings was interested in the ordered and elegant lifestyle of a tiny pond animal called a stentor. A stentor is only one cell big, a trumpet-shaped creature, attached by a 'foot' to a rock on the water bed. It has a tube at its base which can provide shelter, and the trumpet is an open pouch at its free end for feeding. Hairs around the edge of the pouch waft in food particles.

Jennings used carmine, a natural red dye extracted from the cochineal beetle. It can be an irritant even for humans and is certainly toxic to tiny animals like stentor. He added carmine to the water tank in which the stentor was living, and simply watched to see what happened.

The stentor did not at first react to the carmine in the water, but then decisively bent away from the oncoming red specks. The gesture is normally enough to keep it out of trouble in the peaceful conditions at the bottom of a pond. It costs the animal little to try, and it can continue feeding even as it defends itself. In Jennings's experiment, the stentor bent this way three or four times, and when the strategy did not work, demonstrated a second line of defence. It suddenly pushed its pouch out in the opposite direction in an attempt to dislodge any poisonous particles around the mouth. Again this failed, as Jennings continued to drop carmine into the water. Red particles settled on the pouch and a few more similar moves by the stentor proved futile.

Drastic measures were called for, so the stentor retreated. It contracted and moved down into the tube at its base. It waited there for a time but could not wait for ever because a single cell does not store much energy and it cannot feed in its bolthole. It moved tentatively upwards out of the tube, but found the water still full of carmine and had to force itself back down again. It advanced to test the water a couple more times but when conditions had not improved, the tiny creature risked its remaining precious energy, contracted violently, dragged its foot away from the rock and floated away in search of an uncontaminated spot.

Experiments like this have been given new significance by the latest thinking on the adaptive and positive role of fear. A one-celled creature like the stentor has a graded response to a threat, from simply swaying away from the toxin, to pulling up its foot and drifting into the unknown. Stentor allocates its resources meanly so that only the minimum is used to meet a threat. How much more complex, then, might our own reactions to danger be? And could they be built on similar principles?

People do not function in the same way as protozoans, but some of nature's rules are universally true. A great deal of our knowledge of human genetics, for example, is derived from study of the fruit fly, *Drosophila melanogaster*. Our genetic material is the very template from which we grow, and yet most of it can be found in a fly. As Oxford Professor of Physiology Colin Blakemore once rather flippantly pointed out we probably share 70 per cent of our genes with a garden lettuce.

Biologists discovered in 1972 that human cells can apparently commit suicide for the greater good of the whole body. Cells normally receive signals from neighbouring cells telling them to keep going, and should these stop, they die. Cell death is part of the normal development of a foetus in the womb. Babies develop with webbed hands and feet but the skin between the digits normally retreats before they are born. The cells in this skin die, they 'commit suicide' and allow babies to be born with perfectly separated fingers and toes. When cell suicide was first described, it was assumed to be relevant only to the highest creatures since a single-celled organism cannot benefit from its own death. Twenty years after the initial discovery, though, researchers found that single-celled creatures do indeed die in this way. They apparently 'lay down their lives' for the good of their community.

This is just one of many biological similarities between creatures of very different appearance and classification. There is obviously a big difference between the death of a few cells and an all-pervading feeling of fear, but both could be essential for healthy development. Careful observation of animals might help scientists ask more relevant questions about humans. For example, an obvious feature of animals' fear is that it is necessary. If stentor does not react to a toxin, it dies. If the antelope does not run from the lion, it gets eaten. Max Klein lacked normal fear and stood to damage himself socially, financially and physically. All animals need to be able to respond to danger. But how does that help us understand the common phobias?

## Age-Old Anxieties

A mixed bunch of academic publishers, scientific editors and advertising sales staff ate dinner together at the end of a conference. One editor was regaling the table with tales of her previous career as an Avon lady. She lost one of her clients, she said, when she took a swipe at the woman's budgie with her cosmetics bag. Everyone looked up, amazed.

'It was coming straight at me,' she said, by way of explanation. This confident, bright young woman had ornithophobia and was not going to stay in the same room as a free bird.

One of the sales staff was listening with particular interest. 'I know exactly how you feel,' he said with feeling. He was afraid of butterflies and moths, and they started discussing the intricacies of the unpleasantness of wing flapping. Suddenly other diners were vying to compare the strength of their fears. His boss chipped in with a fear of heights and a publisher managed both a fear of spiders and of flying.

The conversation unearthed five phobias in four people among the twelve at the table. Doubtless a psychologist could have found more by interviewing us individually – those mentioned were specific and without much stigma attached – but even this tiny straw poll was telling. The phobias discussed so freely in the restaurant were all directed at threats in the natural world.

No scientist would be impressed by the dubious methods of this survey, but the results are surprisingly reproducible. Whenever a group starts talking about phobias, notice the fears people describe. Occasionally someone has a weird phobia of buttons, cotton wool or wallpaper, and if they do this may dominate the conversation. But most people fear a limited range of creatures or situations. They fear spiders, snakes, the dark, open or closed spaces; creatures and situations that pose few real problems in the West today but which could be dangerous if we lived less cosseted lives.

Evolutionists believe that this observation is important in our understanding of phobias. They say the things we fear today could have been fatal to our prehistoric ancestors. A bite from a spider or snake could have killed; it would have been dangerous to be out after dark; being cornered in a cave by an animal was definitely best avoided. By contrast, the things that really do kill us today – cars, guns or cigarettes – rarely inspire the same level of fear.

They believe that we are, at heart, barely adapted Stone-Agers, now working in offices and driving cars. We are strangely mismatched with our circumstances. We have modern and sophisticated lives but the deep recesses of our mind have developed to react to long-gone situations. The primeval drive of fear is more easily provoked by ancient threats, evolutionists say, because it is still best attuned to days spent roaming the African plains. Then, it would have made sense to have a proper respect for spiders, the dark or enclosed spaces. Stone-Agers lived in dangerous times and required a certain level of caution to survive and have children. Those who did survive passed their safety-consciousness on to their offspring and it became programmed into the human psyche.

The conversation at the dinner table might have ostensibly been about crazy, overblown fears of harmless objects, but an evolutionist would contend that it was in fact about proper caution for dangerous situations – albeit a few tens of thousands of years late.

The theory of evolution has been widely known since Charles Darwin shocked contemporary society with *The Origin of Species*, in 1859. The book had ramifications throughout science, religion and society, as discussed in the previous chapter. It hinted that humankind evolved from a primitive creature over millions of years and is related to the apes. Darwin was initially ridiculed and pilloried for his ideas, but acceptance of them grew and they are now largely taken for granted. In the past decade or so, scientists from many disciplines have revisited evolution theory and attempted to apply it to such diverse questions as why nations go to war and what features of the face or body determine sexual attractiveness. It has been used to argue for a new approach to pest control in agriculture; computers have been programmed to use a kind of technological natural selection to continually improve performance.

But what of our reactions to danger? Can evolution theory tell us anything about the nature of fear and anxiety? Evolutionists claim that part of the reason we develop phobias may lie in the mismatch between life in the twenty-first century and the Stone Age. As a species we are still primed to react to the threats and opportunities that our ancient ancestors faced. Evolutionarily speaking, we have hardly budged in the past ten thousand years but our lifestyle has changed beyond all recognition.

Primates are believed to have appeared sixty-five million years ago, followed thirty million years later by the first ape-like creatures. They began walking on two legs about four million years ago, and using early stone tools two and a half million years ago. After a phase of rapid brain expansion two million years ago, they started to use shaped hand-axes and moved from Africa into Europe and Asia. This was the beginning of the Stone Age and its people developed a stable lifestyle roaming African plains for food until about ten thousand years ago.

Anatomically modern humans developed from their ancestors a hundred thousand years ago and discovered fire. Farming was introduced ten thousand years ago, the wheel about eight thousand years ago, and people started to write about 4000 BC. The pace of change accelerated and it took less than two hundred years to get from the first machines of mass production in the industrial revolution to the technology that put men on the moon.

It is rather like an old man who has lived for seventy years in an isolated spot in an unchanging world. One summer, somebody strikes oil nearby. Big business moves in, a town is developed, new roads are built, the population soars and he finds himself ill-equipped to cope. Humans have spent 99.5 per cent of their existence as hunter-gatherers and are barely out of the Stone Age in evolutionary terms. But life today bears little resemblance to that of our ancestors.

Evolutionists have attempted to explain many modern health problems in terms of the poor fit between our biological make-up and modern lifestyle. Soaring rates of obesity are a good example. Our ancestors had to move around constantly in search of food, which was often in short supply. The ability to store fat around their bodies so that they could survive times of potential starvation would have been a great advantage. Today in most parts of the Western world, food is plentiful. Supermarkets carry a dazzling and expanding range of foods and food shortages are almost unheard of. Add to that a sedentary lifestyle, in which we are entertained at home by the TV, transported around in cars and have our manual work done by machines. The result, according to the World Health Organisation, is that almost half of Britain's adults are overweight and the entire population of America will be obese by 2230 if the increase seen since 1980 continues. Obesity is serious, known to contribute to heart disease, diabetes and premature death. Fat storage, the very mechanism which kept our starving ancestors alive, may be killing people off in the modern world.

The Pima Indians in Arizona, US, are a particularly dramatic example. They maintained a traditional way of life, relying on farming, hunting and fishing for food, until the late nineteenth century. Then, diversion of their water supply by American farmers upstream drove them to poverty and malnutrition, even starvation. The Second World War brought both prosperity and contact with Caucasian Americans, which westernised their dietary and lifestyle habits. Since then, the Pima Indians as a group have put on an unhealthy amount of weight. Half of the adults have diabetes and 95 per cent of those are overweight. Scientists from the US Government's National Institute of Health have studied the Pima Indians for more than thirty years, looking for genetic causes of diabetes and obesity.

Just as the Pima Indians are physically adapted for a traditional lifestyle, our minds may be geared to deal with traditional dangers. We still eat as if food shortages were imminent; perhaps we are also still on the look-out for predators and natural threats. Certainly, evolutionary psychiatrists Randolph Nesse at the University of Michigan, and Isaac Marks, from London University, believe that we are all programmed to react to threats. Anxiety and fear are necessary, they say, and have been essential for our survival throughout evolution.

At the simplest level, mild anxiety boosts performance. It prompts the student to revise for exams, the musician to practise, the sales rep to rehearse a presentation. But evolutionists say it is far more sophisticated than that.

We could improve our understanding of anxiety at a stroke if we stopped thinking of it as a disorder, and considered it a defence that regulates and orchestrates our reactions to every threat and opportunity, say Nesse and Marks. The anxiety system is as important to our survival as is our immune system. It protects against threats to our whole body, and life, in the way that the immune system fights off specific physical threats. Both defence systems have developed within our species as we evolved. The individuals with appropriate reactions to danger or to micro-organisms are most likely to survive, produce off-spring and pass on these traits to future generations.

Both have a range of reactions to meet specific threats. The immune system creates a scab to heal a cut finger and produces antibodies to deal with viruses. Similarly, at least some of our reactions to danger are clearly adaptive and matched exactly to the threat, say Marks and Nesse. For example, people who are afraid of heights may 'freeze' if they have to walk along a ledge or cross a narrow bridge. They cling to the side, unable to move. They may need a companion's reassurance and physical assistance to get going again. This sort of reaction is not helpful if it stops you climbing stairs but in natural surroundings someone who became immobile by the side of a sheer drop might avoid a bad fall.

Blushing often seems to make a difficult situation worse. People who lack confidence in social gatherings dread being the centre of attention and burning cheeks do not help anyone blend into the background. But if, as has been argued, blushing signals social submission, a red face could be a plea for continued membership of the group. In ancient times, membership of a group would have been near-essential for survival. Anyone expelled and left alone would become vulnerable to cold, starvation and attack. They would also be unlikely to pass their influence on to the next generation if they could not reproduce.

Blood and injury phobias provide an intriguing physiological example of the possible benefits of an anxiety response. People with these phobias may faint at the scene of an accident or even at the sight of a syringe or needle. These are the only phobias associated with fainting. Someone with agoraphobia may feel extremely dizzy or uncomfortable in a crowded street and believe they are going to pass out, but they almost never do. As the agoraphobic prepares to flee the difficult situation, rising blood pressure effectively prevents a faint. By contrast, the blood and injury phobics' blood pressure drops at the sight of blood, and they often do pass out. Nesse and Marks argue that this, again, could be adaptive. If a hunter saw blood, it was more likely to be his own than anyone else's. An injured man loses less blood if his blood pressure drops. Even if he fainted, this could conceivably be useful. Some animals only attack moving creatures and lying still might just discourage further attack by predators.

Many animals are known to play dead while remaining conscious. Charles Darwin himself once caught a robin in a room and said it 'fainted so completely, that for a time I thought it was dead'. David Barlow, an eminent psychologist in Albany, New York, says there may be a human parallel. Women who have been raped frequently describe being paralysed, rigid and cold during the attack. They are not unconscious because they can later remember details of what happened. In the past, this freezing has been wrongly taken by courts to mean that the women somehow consented to sex. Barlow says their immobility may in fact be an ancient defence mechanism. Remaining still may reduce further violence by a more powerful assailant and could conceivably reduce his sexual arousal.

In this way, the nature of a reaction is matched to the threat. Blushing is not likely to scare off a snake and freezing would not help in a difficult social situation. Normal phases of development also fit the evolutionists' model. Babies may suddenly become afraid of strangers between six and twelve months old, just when they are starting to crawl and coming into more contact with unknown people. Animal fears peak at about four years old, the age when they may start meeting and playing with animals unattended. Social phobia typically starts in the late teens, just when young people are establishing their identities and facing all sorts of social pitfalls. While it would be unwise to take the argument too far – even Marks and Nesse have admitted that imaginative thinkers could come up with an adaptive use for virtually any human reaction – there are many compelling examples.

The strength of a reaction to a threat is every bit as important as its nature. Both the anxiety and immune systems are tightly regulated and over- or under-reaction causes problems. The human immunodeficiency virus, HIV, does not itself kill, but its destruction of immune defences means normally harmless bacterial and viral infections can become fatal. At the other end of the scale, allergies and hay fever develop when the immune system is overreacting to irrelevant stimuli like dust or pollen.

Anxiety is similar, argue Marks and Nesse. An underactive anxiety system may create real problems, as demonstrated by Max Klein in *Fearless*. A lack of concern about the future sounds wonderful, but not if this destroys all ability to plan for it. Never worrying about the consequences of your actions may mean you speak out when it would be diplomatic to say nothing. Telling your boss exactly what you think of him or her is a fantasy for many of us, but we never do it. A moment of extreme satisfaction could cost you your job. Similarly, you might feel like objecting loudly when someone pushes past you at a bar, but if they are big, drunk and bad-tempered, you probably keep your feelings to yourself. Those without normal levels of anxiety may lack basic caution and end up losing jobs and getting into fights where others simply sidestep trouble. Without the push of anxiety, it may be difficult to revise for exams or apply yourself to any long-term project. Marks has termed this hypophobia. It is interesting but speculative. It has not been studied much because those who lack anxiety often don't imagine they have a problem and tend not to come forward for help. However, New Zealand researchers have some evidence to back the idea and at the same time, challenge the widespread assumption that a traumatic experience can trigger a phobia. They looked for height phobias among children who had serious falls between the ages of five and nine. They found that, at eighteen, these children were much less – not much more – likely than others to have height phobias. This study implies that temperament (discussed in chapter 7) may be all-important and suggests that children without fear, those who have never worried about heights may be hypophobic, and most likely to injure themselves in a fall.

The over-reactive end of anxiety is far more familiar. A wealth of anxiety disorders, including phobias, result directly from a tremendously sensitive anxiety system. People with these disorders can become upset by things others would never notice. Hoarders, obsessives and agoraphobics fear things but they all have hair-trigger anxiety systems. The hoarder is so afraid of losing something important that he cannot throw away anything. His house gradually silts up with layers of junk and old newspapers. The obsessive washes and cleans for three hours every morning and is quite unable to go to work unless she, and the house, are immaculate. The agoraphobic may hear about a road accident fifty miles away and be housebound for days afterwards.

Nesse carried out an interesting exercise in which he listed the physical and social dangers that would have threatened early humans. Physical dangers included accidents, disease, starvation, predators, hostile humans; social dangers included rejection, attacks on status or disruption of relationships. Modern anxiety disorders correspond well with these ancient threats. The hunter-gatherer's proper fear of predators could have become today's animal phobia; storage of food in times of plenty to ward off starvation could have become hoarding; cleaning rituals and taboos to ward off disease or contamination could have become obsessive-compulsive disorder. The hunter who sensibly stayed at the home base while a hungry lion roamed may have become today's agoraphobic, highly reluctant to go out.

Responses that may once have been life-saving reactions have become inappropriate. Fear of heights, once a proper respect for the danger of a high cliff, is now a nuisance if it translates into fear of bridges or high-rise apartments. Reluctance to approach spiders may have been wise, and still is in some parts of the world. But fear of spiders in countries like Britain, where none is harmful, is widespread, and serves no useful purpose.

Nesse's point is that today's anxiety reactions would often have been essential in the Stone Age. There is nothing essentially wrong with the reactions, they are just too easily triggered for life today. It is a helpful idea. Fear of danger is a natural response and one which in other circumstances, thousands of years ago, might have protected us rather than blighted our lives.

## The Evolution of Fear

The idea that we are attuned to life on the African plains makes a wonderful story, but most of us do not *feel* much like Stone-Agers. We have adapted to many changes even in the last few decades; we are more or less at ease with cars and aeroplanes, computers, dishwashers and foreign holidays. How come our fears lag so far behind?

Can we really blame our prehistoric ancestors for our fear of snakes and spiders? Fear may be contagious but evolution demands that it is passed down for tens of thousands of years. It is asking rather a lot for fear to survive intact so long. Diseases have died out in that time, whole species have become extinct. Yet evolutionists say that our fears of heights and the dark have remained unchanged since our predecessors in the Stone Age were trying to get back to their caves at night.

Charles Darwin first introduced evolution to the public in the mid nineteenth century. Some of his basic ideas were old, even then; Charles's grandfather, Erasmus Darwin, had been one of several advocates of the theory in the eighteenth century. Charles Darwin himself became convinced during a five-year voyage through the southern hemisphere on *HMS Beagle*. He watched species of animals change gradually from island to island as the boat moved south. But it was his observations in the Galapagos Islands that were critical to forming his theories. Finches and giant tortoises varied slightly but predictably from island to island and the local people could always tell where a particular bird or tortoise belonged. It seemed that all the variations of these creatures must have had a common ancestor.

It took Darwin twenty years to publish *The Origin of Species*, possibly in part because he anticipated and dreaded the uproar it would cause. The book is packed with examples of evolving creatures: ants and bees, horses and zebras, birds, fish and plants. Darwin defined the process of natural selection as 'the slow and gradual accumulation of numerous, slight, yet profitable variations'. If a tiny random change in an individual gives it an advantage over other members of the species it is more likely to survive, reproduce and pass the adaptation on to its offspring. The offspring in turn are more successful than those without the adaptation and over many generations more and more of the population are born with this small advantage. In the same way, variations which give their carriers a disadvantage are eliminated.

Natural selection works to produce gradations in animals' instincts as well as in their physical features, Darwin said. Nesting birds, for instance, have an instinctive fear of most of their enemies, strengthened by their own experience and by the sight of fear in other birds. But they are slow to develop a fear of humans. According to Darwin, large birds in highly populated countries like England are wilder than small birds because they have been persecuted by humans. In uninhabited islands, large birds have no more fear than small birds. Magpies and hooded crows are wary in England but, in Darwin's time at least, magpies were tame in Norway as were hooded crows in Egypt.

Darwin omitted humans from his arguments in *The Origin of Species*, but discussed them at length in later books, *The Descent of Man*, 1871, and *The Expression of the Emotions in Man and Animals*, 1872. *Expression of the Emotions* set out to demonstrate that different races, and even different species, show their emotions in a remarkably similar way, implying that emotions such as fear have been conserved throughout evolution:

That the chief expressive actions, exhibited by man and by the lower animals, are now innate or inherited – that is, have not been learnt by the individual – is

admitted by every one. So little has learning or imitation
to do with several of them that they are from the earliest
days and throughout life quite beyond our control: for
instance, the relaxation of the arteries of the skin in
blushing, and the increased action of the heart in anger.

Fear is one of these chief emotions:

Fear was expressed from an extremely remote period,
in almost the same manner as it now is by man; namely,
by trembling, the erection of the hair, cold perspiration,
pallor, widely opened eyes, the relaxation of most of the
muscles, and by the whole body cowering downwards or
held motionless.

Or, as he says elsewhere, quoting from Papinus Statius'
*Thebaid*: '*Obstupui, steteruntque comae, et vox haesit*' ('I was
astounded, my hair stood on end, and my voice choked in
my throat').

Erection of the hair is singled out for special comment
because it serves no purpose in humans and may simply be
a leftover from evolution.

With respect to the involuntary bristling of the hair, we
have good reason to believe that in the case of animals
this action, however it may have originated, serves,
together with certain voluntary movements, to make
them appear terrible to their enemies; and as the same
involuntary and voluntary actions are performed by
animals nearly related to man, we are led to believe that
man has retained through inheritance a relic of them,
now become useless. It is certainly a remarkable fact,

that the minute unstriped muscles, by which the hairs thinly scattered over man's almost naked body are erected, should have been preserved to the present day; and that they should still contract under the same emotions, namely, terror and rage, which cause the hairs to stand on end in the lower members of the Order to which man belongs.

Darwin therefore argued that the similarity of our and animals' response to danger is further proof of our common ancestry. But not all emotions are so ancient. Other emotions, such as blushing through shame, shyness or excessive attention, have developed more recently, he said. Different races of people across the world all blush, but animals never do.

It does not seem possible that any animal, until its mental powers had been developed to an equal or nearly equal degree with those of man, would have closely considered and been sensitive about its own personal appearance. Therefore we may conclude that blushing originated at a very late period in the long line of our descent.

Or as Mark Twain wrote, 'Man is the only animal that blushes. Or needs to.'

Darwin referred to innate and inherited fears, but a key problem with his argument at the time was that he could not explain how physical or emotional traits were passed down. In fact, the Moravian monk Gregor Mendel was coming up with answers even then. Mendel was working with peas and developing the idea of genetic traits conserved through generations. This is now generally accepted as the mechanism by which evolution works and the idea of genes being passed from parents to children, determining family traits and peculiarities, is a familiar one.

Genes are the template from which we develop. They influence all aspects of our physical and mental well-being, including our appearance and vulnerability to diseases, our intelligence and personality. They are not the whole story and our environment also plays its part. But genes certainly influence the development of the brain's structure and the activity of chemical messengers involved in our experience of fear. As explored in more depth in the next chapter, our genes could programme our brains to react to danger.

Although it may have taken millions of years for the lumbering progress of natural selection to give the world its incredible diversity of species, Nesse points out that individual traits can change much more quickly. Selective breeding of dogs for their temperament, for example, takes just a handful of generations to produce puppies which are either exceptionally easygoing or frantically neurotic.

But genes may not be our only link to the Stone Age. A chain of people also exists. If we are lucky, we know our parents and their parents and we may be familiar with the two or three, possibly even four generations which preceded us. These generations in turn knew their parents and grand-parents and so on back through history. The human links are continuous and it could be that we learn our behaviour from those around us, just as they learnt it from another generation. A kind of cultural rather than genetic trans-mission of fear could take place.

Humans have passed on information since we could paint on cave walls and tell stories. Today's films, books and TV programmes may be efficient ways of relaying a fear of ancient threats. When a writer wants to create a threatening atmosphere, a dark night, a few large spiders and a pair of animal's eyes usually do the trick. Many ancient threats are symbols or shortcuts to fear and the storyteller only has to mention them to create the desired mood. Cultural learning is powerful and flexible and can quickly shape attitudes in an enduring way.

The idea was proposed by psychologist Graham Davey from the University of Sussex, and he set out to test it in relation to fear of spiders. Spiders have been embroiled in European culture since the Middle Ages, when they were thought to absorb poisons and to infect any food they touched. They were seen as the forerunners of disease and death during the Great Plague (the discovery that rats' fleas carried the disease was not made until the nineteenth century). A form of hysteria called tarantulism was even blamed on the spider and only later found to be caused by too much sun. It was not all bad for the spider – tiny 'money spiders' were thought to bring financial good luck and cobwebs were used in traditional medicine to bind wounds – but for hundreds of years, in the main, spiders were thought to be highly dangerous and widely feared.

This thinking was confined to Europe, so if cultural learning is responsible for fear of spiders, Davey reasoned that it should only be widespread in Europeans and their descendants. Observations back him up. In parts of Africa the spider is thought to be wise and local people clean and protect its habitat. Spiders are eaten as a delicacy in areas as diverse as Indo-China, the Caribbean, among the native North Americans and Australian aborigines. Children in Brazil often keep spiders as pets. Hindus in eastern Bengal collect spiders to release at weddings to wish the couple good luck, and in

Egypt it is common to put a spider in the bed of a newly married couple.

Researchers found that the incidence of spider fear in Britain is similar to that in Holland and in the US. So far, so good. Many North Americans are descended from Europeans, so this is not unexpected. The incidence in these countries was higher than in India, as predicted. But, strangely, the incidence in Japan, where there is no particular history of spider fear, is even higher, which tends to weaken the argument.

Davey's central point is that it takes countless generations for the biology of a population to change even slightly. Threats would have had to be extremely dangerous and common, killing people in large numbers, for fear reactions to have become biologically programmed. He says we must not ignore the costs of our reactions. Our ancestors might have been well-advised to keep away from poisonous spiders or snakes, but they had to grub through plants to get food. Too much fear of insects would have led to malnutrition if it made people reluctant to look for food. An infant starting to explore its surroundings might be at risk from strangers and suspicion might be appropriate. But strangers are also likely to help a child in trouble and over-reluctance to approach a stranger could be fatal.

Spiders and snakes may simply have had longer to become embroiled in our culture and inherited learning than modern threats. Guns and electricity outlets have not been around long enough to acquire the symbolic significance that would mark them out as objects to fear. Children may develop fears by absorbing information from the people around them, who are themselves more likely to fear snakes than guns.

Cultural transmission of fear was a bold challenge to the

prevailing view that our thought processes are shaped by strong biological links with our ancient predecessors. It suggested that our fears may have nothing to do with our biology and that perhaps our primitive brain was not, after all, programmed over millions of years. It is possible that we have learned them solely through careful observation of those around us.

However, this idea has not caught on. Nobody denies the importance of learning, but some of the most exciting research work is attempting to examine the structure and activity in our brains. It seems most likely that genetic and cultural transmission of information work in tandem. We have evolved with a certain biological background which comes to life only in the context of cultural learning. The tendency to fear may be instinctive or hardwired, irrevocably programmed into us as a species. But personal experience and observation of others may be essential before we develop specific fears.

The idea of flexible learning overlying hardwired fear has been re-explored by evolutionists in recent years. They are delving into aspects of the theory and attempting to test them out in practical, modern ways.

## Animal Instincts

One of the good things about being an evolutionist is that you can never be definitively contradicted. Most scientists have their best work overturned within their own working life. They spend time trying to disprove other scientists' ideas but they in turn are usually overtaken by someone else who contradicts or at least refines their work.

Believing in evolution gives a scientist some respite. Evolution took place over such a phenomenally long time scale that we can never recreate the same conditions and, ultimately, never know anything for certain. It provides a rather luxurious and permanent platform for scientists to stand on.

This does not mean that we have to accept the evolutionary perspective without question. With some lateral thinking, many ideas stemming from evolution theory can be studied scientifically. For example, evolutionists say that we are more likely to fear ancient rather than modern threats. If this is so, it should hold true for people of different races and cultures since we share the same ancestry and should therefore share the same fear programming.

A few small studies have produced some evidence for this. Researchers at a mental health clinic in Bangalore, India, found an incidence of phobias only a tenth of that in the West, a rate similar to that in other Indian communities. However, the vast majority of the phobias fitted the evolutionists' model. Agoraphobia was the most common, closely followed by illness and social phobias. Animal phobias were rare, which is usually the case in clinics catering for people with the most seriously disabling problems. Scottish work found that more than two-thirds of a group's phobias were relevant to ancient times. A Sri Lankan study used the same method and came up with virtually identical figures. This provides some backing for the idea that people in different parts of the world are similarly attuned to fear threats in the natural world.

More fundamentally, we cannot apply evolution theory to phobias at all unless we think cautious Stone-Agers were more likely to survive and produce offspring than their fearless friends. Fearfulness should have increased the chances of people passing on their genes to the next generation.

Coupled with this is the demand that ancient threats which still provoke fear today were capable of killing off people in large numbers, or at least reducing their chances of having children.

We cannot easily test this out in humans, but we can look for evidence in animals. Darwin noticed that birds are more ready to fear cats than people, presumably because they are at more risk from cats. More than one hundred years after Darwin, a psychologist at the University of Pennsylvania, Martin Seligman, became interested in how that should be. After all, both cats and people kill birds and it might be as well for the average bird to have a healthy respect for both.

Seligman said that birds are programmed to fear cats but not people. Somewhere in the depths of the bird's tiny brain lies the knowledge or the instinct that makes them ready to fear cats. By contrast, they are essentially neutral towards humans. Birds may become afraid of humans, but are not likely to fear people unless they have been harmed or hounded in some way. Cats are natural enemies of birds and have killed off swathes of them down the ages.

Seligman said that birds are 'prepared' to fear cats but 'unprepared' to fear people. He said further that some animals are 'contra-prepared' to develop certain fears and never become afraid even if they have repeated bad experiences. For example, pigeons instinctively peck for food and in the laboratory they learn quickly to peck a lighted key if it delivers grain. But if the experiment is set up so that pecking the key *prevents* them getting grain, they do not learn, and continue pecking at the key even though they never get anything to eat. Pigeons normally have to peck to feed and they are contra-prepared to make an association between pecking and starvation. The hungrier they are, the harder they peck, and it never occurs to them that taking a rest might be the answer.

Similarly, they learn quickly to fly away to avoid a shock but only with great difficulty to peck a key to stop the shock. Again, it makes sense. Hopping or flying away from an unpleasant stimulus is a good idea. Pecking, as a rule, would not help.

Humans may also have degrees of preparedness to develop fears. Watson and Rayner's experiment with Little Albert (described in the last chapter), showed that he learnt instantly to fear the furry rat, and many other similar objects, after the experimenters startled him with a loud noise while he was playing. He did not take against the scientists conducting the work, who quite clearly deserved it, which suggests he was more ready to fear animals than people. Another researcher gave children common household objects like curtains and blocks to play with, delivered a sudden loud noise, and found they developed no fear at all. In yet another similar experiment, children remained robustly unafraid of a wooden duck.

This may be an important hint as to how phobias develop. The brain's hardwiring determines how ready we are to become afraid of something. It provides a kind of mould for our fears. Some animals and situations fit it well. Fears of them are instinctive and develop with the least provocation. Seligman initially said that objects or situations which threatened the survival of the species, such as insects, animals, heights or the dark, best fit the mould. Phobias develop without conscious thought, sometimes after a one-off event, and they are not easily extinguished. In addition, the more flexible process of experience, learning and observing others, means that we can become afraid of anything. Modern objects tend not to fit the mould and it takes very adverse circumstances to make us fear them. Phobias of objects can and do develop, but not easily. Seligman rather harshly suggested that people with these fears may have 'talked' themselves into it. Fear of cars or guns is unprepared, he says, our brains have no template for it, and it takes effort or severe experience to lodge the fear in our minds. We will return to this point later in the chapter.

For many years there was a fierce debate over whether people and animals were born with fears or developed them later, but in the 1960s researchers demonstrated that laboratory-reared monkeys are not at all afraid of snakes. Monkeys who have spent even a short period in the wild are extremely afraid. It is highly unlikely that all of the once-wild monkeys had had a traumatic experience with a snake, so this was puzzling and it seemed that at least some of the monkeys must have acquired their fear vicariously, through seeing another monkey acting scared – a kind of fear by proxy.

Susan Mineka and colleagues at the University of Wisconsin put a monkey's favourite treat, such as a marshmallow or raisin, on a ledge behind a transparent box. There was a real or toy snake in the box and the monkey had to reach over the snake to get the sweet. The more afraid the monkey was, the more reluctant it was to stretch over the box, and Mineka found that the fearless laboratory-reared animals grabbed the treat where the once-wild monkeys refused it.

Some laboratory-reared monkeys had lived with their previously wild parents all their lives. This was obviously not sufficient for them to acquire the fear of snakes – they seemed to need some experience for the fear to develop. When the laboratory-reared monkeys were allowed to watch older, wild-reared monkeys cowering from the snake, the vast majority developed the same reaction themselves, within minutes. They mimicked the screen monkey's behaviour, clutching or shaking the cage, grimacing or threatening.

Mineka then attempted to make monkeys fear flowers or rabbits, objects that could never pose a threat. One video showed a monkey afraid of a snake and another was edited so that the same monkey was apparently afraid of a flower. Fearless young monkeys watched the tapes and afterwards, those who had seen the snake tape avoided snakes, but those who had seen the flower tape remained unconcerned by flowers. It was clearly easier to induce a fear of snakes than flowers. A similar experiment demonstrated that the monkeys were more ready to fear toy crocodiles than toy rabbits.

This is powerful evidence that creatures really are programmed to fear certain things. These monkeys were born in laboratories and had never previously encountered snakes, flowers, crocodiles or rabbits. Mineka and Cook concluded that it was highly likely that the difference in the monkeys' reactions was somehow in-built, or 'phylogenetic'. In other words, snakes and toy crocodiles are a better fit for the mould in the monkey's brain.

Monkeys may not be born afraid of snakes but any sort of demonstration is enough to provoke their fear. It makes sense from the evolutionists' point of view. Animals may not get a second chance in the wild and mistakes can be fatal. It could be that monkeys that quickly learn to be afraid of snakes or crocodiles have a survival advantage over their bolder companions. They are more likely to avoid these animals and therefore to survive and produce offspring. They will pass on the tendency to fear and, over generations, natural selection would increase the proportion of all monkeys inclined to fear snakes or crocodiles.

Work like this could not be done on people because we would all have experience of any object the researchers chose.

But it might still be possible to draw human parallels from the work. Monkeys are not people, but our learning processes are surprisingly similar.

Take a country like Britain, where we have only one poisonous snake, the adder, and virtually none of us has ever seen it. Yet many of us are afraid of snakes. Why? Mineka's work suggests, if humans are anything like primates, it will not take much exposure to snakes for a strong fear to develop. Monkeys developed permanent fears from watching videos and people probably do, too. We could be watching from a distance as someone else reacts to a snake or, much more likely, see someone shuddering at them in a film or on TV. From a very young age, we learn of Little Miss Muffett being frightened away by the spider, or the farmer's wife shrieking in terror at three blind mice.

Role models are powerful, especially – according to Mineka – if they are older and more dominant. Her models did not have to be related to the young monkeys but it helped if they knew each other. This suggests that parents or other influential adults – even television and film role models – could pass on their fear to children. Mineka believes that if adults have phobias, they should not confront snakes, spiders or whatever it is they fear in front of children. We might expect it to be a bad thing for parents to blatantly avoid objects or situations, but this study suggests that it is worse for children to see their parents visibly disturbed.

There is a plus side to this work. Mineka found that monkeys can be immunised against developing a fear and learn *not* to be afraid. A model monkey who was unafraid of a snake made a lasting impression on the naive monkeys. They apparently got the message that snakes are not to be feared and it prevented the later development of fear. If, afterwards, they saw another monkey afraid of the snake, three-quarters of the young monkeys remained fearless. This suggests that adults who show no signs of fear when dealing with spiders or snakes exert a powerful influence on children and may prevent them developing these fears.

Mineka's work tells a neat story, based around the assumption that snakes and crocodiles were real threats to monkeys and killed them off in huge numbers. Monkeys who were afraid of these animals therefore had a survival advantage. Unfortunately, the evidence does not fully back up the theory. The rhesus monkeys used by Mineka evolved in India, where cobras and other poisonous snakes could have been dangerous. However, there is less evidence that crocodiles would have been a danger. Crocodiles might be feared because of their reptilian similarity to snakes, but this seems rather to weaken the argument.

It partly hangs on how the brain recognises threatening animals or situations. The brain could have a full picture of snake or crocodile irrevocably programmed into its hardwiring. Alternatively, features like smell, sliminess or sudden movements may be what we are on our guard for.

People with phobias often give vivid descriptions of what they fear; the appearance, feel or thought of the animal. Abrupt, jerky, unpredictable movements are frightening. Sliminess disgusts us. Even babies dislike strange, inhuman appearances. Jamie Bennet-Levy and Theresa Marteau in London asked a group of people about rats, cockroaches, butterflies, frogs, rabbits, spiders, blackbirds and other small animals. The volunteers rated each creature for ugliness, sliminess, speed and how suddenly they appear to move. Another group said how afraid they were of the various characteristics and how near they would go to each animal.

Not surprisingly, the more harmful the animal was, the more afraid people were and the less prepared to get close. Physical characteristics, especially ugliness, also deterred them. The volunteers in the study said that ugliness was a composition of sliminess, hairiness, colour, dirtiness, number of limbs or antennae, compactness of body and the relation of the eyes to the head. In other words, how different the animals' appearances were from humans. Touch and sound came into it as well and people hated the thought of a spider running up their leg or in their hair.

Similar work in the Netherlands also concluded that the more animals differ from humans, the more we fear and avoid them. Lack of predictability or control of the animals makes things worse. People are inclined to be suspicious of all things strange and seem ready to fear features of alien species. This suggests that the human brain is set up to notice unpleasant characteristics and probably does not contain a full picture of certain dangerous animals.

It is puzzling, then, that Mineka found monkeys quick to fear toy snakes and crocodiles, which do not have strange smells or movements. Suffice it to say that not even evolutionists can claim that all evidence points the same way.

The inner workings of the brain remain mysterious, despite increasingly sophisticated work such as that described in chapter 4, on neurophysiology. Tracking down the unconscious requires ingenuity and a Swedish group led by Arne Öhman and Joaquim Soares hit upon the idea of using subliminal images. The approach has a certain history. A classic but flawed experiment in the 1950s flashed up a message urging cinema viewers to drink Coca-Cola. The message disappeared so quickly that viewers had no idea their film had been interrupted, but this brief exposure was enough to make them inclined to buy more of the drink. However, the film itself was about food and drink and the results were inconclusive. In any case, sadly for the advertising industry, such tactics remain illegal.

In the laboratory, Öhman and Soares had more freedom and flashed pictures of spiders, snakes, flowers and mushrooms on screen. They recorded skin conductance on volunteers' hands. It is a measure of arousal in the autonomic system and changes when people have any sort of anxiety reaction. Volunteers responded to the subliminal pictures. For example, people with spider phobias reacted more strongly to pictures of spiders than anyone else. They did not know what they had seen but nevertheless had a physical reaction. Like the cinemagoers who suddenly had an urge to drink cola, the volunteers unconsciously registered the picture they had seen.

The researchers then tried to make the volunteers artificially phobic and gave their finger an unpleasant electric shock as the flashed pictures appeared. Some received the shock when they saw the snake picture, others when they saw the spider, flower or mushroom pictures. They all reacted to the picture paired with the shock.

Volunteers then saw the flashed pictures again but without the shock. Responses to flowers and mushrooms were eliminated but responses to snakes and spiders endured. None of the students had any conscious idea what they were reacting to, but responses to snakes and spiders were consistently more resilient than those to flowers and mushrooms. It implies that the unconscious can somehow pick out fearful stimuli.

By the time we recognise a spider or snake, our unconscious is already generating a rising tide of arousal which puts us on red alert, said Öhman and Soares. This heightened awareness makes us more likely to feel defensive and afraid.

Phobics cannot control their fear voluntarily because the initial reaction is unconscious and under the control of deeper, ancient parts of the brain.

These exciting results encouraged the researchers to look at another type of danger: threatening human faces. Threats from predators such as snakes are important to the most primitive creatures and defences would have evolved early. Social threats affect more sophisticated creatures and would have evolved much later. Therefore the relevant information may be processed in the higher brain, the cortex, which carries out more complex functions and developed late in evolution. This ties in neatly with Darwin's earlier assertion that blushing developed comparatively recently.

The cortex has left and right halves with different functions. If social submissiveness is processed in the cortex, our reaction may depend on where the human face is when we see it. Negative emotion and perception of faces are believed to be processed in the right hemisphere in right-handed people. Information from the left side of someone's visual field is also processed in the right hemisphere. If someone sees an angry face in their left visual field and the information goes straight to the hemisphere adapted to process it, they may react more vigorously than if the information is sent to the other side of the brain. There are lots of ifs here, but volunteers did react more strongly to a flash picture of an angry face when it was shown to their left rather than right visual field.

If fear of small animals is processed by the oldest parts of the brain, deep down where there is little difference between right and left sides, there ought to be little difference whether the slides are shown to the right or left visual field. This turned out to be true. People's reactions were similar, regardless of where the slides were flashed up, which supports the idea that this reaction developed early in evolution.

Information on ancient threats is processed first by ancient parts of the brain, according to Öhman. Colour, texture, smell or type of movement may be sufficient to trigger this automatic response and we react immediately and instinctively. We may even have special pathways in the brain for transmitting information about ancient threats, pathways which existed before sophisticated reason and logic and which are physically distinct from those carrying conscious thought. This ancient reaction can produce a groundswell of emotion which colours our thoughts even before the higher brain has engaged and started to make us aware of a potential threat.

Like Seligman, Öhman suggests that responses to ancient threats are easier to establish and harder to extinguish than responses to unnatural or modern cues. It certainly provides a neat explanation for the reactions that may underpin many phobias. The reality is, unfortunately, less straightforward and Öhman's results have proved fragile. There are many detractors, and even some supporters of the theory have been unable to come up with the same answers.

An American group led by Edwin Cook, himself an advocate of evolution theory, set out to repeat some of Öhman's work. They used a similar set-up and were surprised to find that reactions to spiders and snakes were as easy to induce as modern fears of guns. It was also equally easy to eliminate them. In direct contrast to Öhman's predictions, reactions to spiders disappeared as quickly as reactions to guns.

Cook used an unpleasant noise rather than an electric shock, so his volunteers received no tactile stimulus. When he added a vibratory stimulus to the hand, his experiments started to distinguish between modern and ancient threats, but the results were unconvincing. There was still no difference in acquisition or extinction of fears, though Cook found an increased heart rate associated with spiders or snakes and not guns.

When Cook followed Öhman's protocol to the letter, he confirmed that fear of ancient objects was more resilient than that of modern threats. But the difference was slight. Overall, Cook's work provided only flimsy support for Öhman and also raised questions about the relevance of some features of the experiment.

Cook was not the only one to have trouble replicating Öhman's work. An American duo, Richard McNally from Chicago medical school and Edna Foa from Pennsylvania, changed mushrooms for strawberries in Öhman's experiments. This was enough to eliminate differences in reactions to threatening and non-threatening objects. Even people who were phobic of snakes or spiders before the experiment reacted to strawberries just as strongly.

Perhaps the greatest disappointment to advocates of the theory is that it has little impact in the phobia clinic. Despite the predictions of Seligman and Öhman that fears of ancient threats are more easily acquired, more deeply ingrained and harder to reverse, this seems not to be the case. The Scottish and Sri Lankan studies mentioned earlier both found that the vast majority of phobias in these two quite different countries are related to ancient threats. However, neither study linked phobias of ancient threats to more severe impairment. Zafiropoulou, who led the Scottish study, said that success of treatment was unrelated to the type of phobia and suggested that the concept of phobias as vestiges of our ancient past is little practical use.

Evolution theory is intellectually pleasing and is enjoying a well-deserved resurgence of interest. That does not make it true and it is near impossible to prove or disprove the theory. Evolutionary biologist George Williams was asked on a radio programme whether people are still evolving. Yes, we are, he said, at a fantastic rate, and thirty or forty thousand years from now you'll really see the difference.

Many of the questions raised by the theory have not been satisfactorily answered. The experimental evidence is limited and far from conclusive. Brain scanning and other future work may determine which parts of the brain are involved in our emotions and reactions, but we will never know for certain why that is.

We do at least know that phobias of all types can be successfully treated. Perhaps the most useful application of evolution theory to people with phobias will be to help those with seemingly ridiculous fears to understand and come to terms with them. Evolution can provide a framework for examining deep-rooted anxiety, whether or not it has anything to do with our ancestors' lifestyles. Most reassuringly, evolution works in tandem with the more familiar sciences. It is consistent with the influence of genetics and physiology, it contradicts neither cognitive, behavioural nor even psychoanalytical approaches for treating phobias. We can choose whether or not to believe that our brains were shaped by evolution, but that need not halt the search for more information on what triggers anxiety today and how to deal with it when it becomes unbearable.

## CHAPTER 3

# Genetics

## Icy Fear

Humans, like rats, have the convenient ability to survive in all climates. Whether tropical, temperate, barren desert or icy bleak, we manage to live in most areas of the world. We build our homes on steep mountainsides and on vast plains, by coasts and in humid tropical forests. We are a remarkably adaptable species.

This hardly seems a feat; living as we do in comfortable centrally heated homes, with wardrobes full of clothes for different seasons, it is not something we need to think about often. But the fact that people populated these diverse areas long before they had any of today's mod cons may, believe it or not, relate to phobias.

Our predecessors lived in Africa for most of the last four million years. They had started to leave Africa and move north into Europe and Asia by about one hundred thousand years ago. Some went only as far as southern Europe, but others ventured further and 30 to 40,000 years ago began making their homes in the freezing lands of the north. The climate in northern Europe, severe even now, was extreme at this time before the glaciers of the Ice Age had retreated.

These people were presumably adapted for the African

climate, but survived in the harsh temperatures. They found food and shelter, they settled down and had families. Somehow they thrived in the intensely cold environment.

Jerome Kagan, psychologist at Harvard University, was intrigued by their success and came to wonder whether they had undergone a physical adaptation to help them cope with the climate. They could have grown thick body hair or developed rolls of fat to insulate themselves against the cold. But northern Europeans are typically tall, slim and fair, not fat and hairy. Kagan believes they may instead have had raised levels of a chemical messenger which stoked up their metabolism and increased body temperature. According to Kagan's fascinating, if somewhat convoluted, hypothesis, a single genetic mutation may have altered physical appearance and anxiety levels. It may sound unlikely, but the logic is compelling.

The chemical messenger is noradrenaline. Ancient Swedes and Finns could have benefited from raised levels of noradrenaline, which raises body temperature and shuts down the skin's blood supply, reducing heat loss. Anyone lucky enough to have the slight genetic variation which raised noradrenaline levels and kept up their core temperature would have had a real advantage in a climate cold enough to kill large numbers of people. They would have been more likely to survive in freezing temperatures, to function normally and be able to mate and pass the variation on to their offspring. Thus over generations the variation would have become common.

Noradrenaline has many functions throughout the body and acts on ancient centres of the brain that adjust our thresholds for fear. Higher levels of noradrenaline can mean higher levels of fearfulness. The northern European's temperature rise might have been won at a cost if it made people with warmer bodies more prone to anxiety. In this environment, the ability to resist the cold would have been the overriding factor, and fearfulness a price worth paying.

What this lovely tale needs is direct evidence that northern Europeans are more anxious than other races. No such data exist, but Kagan's belief that the theory offers insights into the genetics and mechanisms underpinning anxiety has driven him to gather diverse observations to back it up.

Kagan believes that the same mutation which altered nor-adrenaline levels may also have influenced physical appearance. Northern Europeans classically have blue eyes, fair complexion and a slim build. He says that this general appearance is popularly associated with nervousness and, further, that anxious children frequently share these characteristics. Film directors, always on the look-out for visual short-cuts to describe a character, are unlikely to cast a slim, blond, blue-eyed actor in the role of school bully. The southern European's typically dark eyes and stockier build is associated with a relaxed, confident attitude to life and again this is seen in films. Sylvester Stallone does not look like a man who gets bullied. Even cartoon characters' appearance is in keeping with their fearfulness levels. A colleague of Kagan's, Doreen Arcus, looked at seven major Walt Disney films and found that fearful characters like Cinderella were more likely to have blue eyes than bolder ones like Peter Pan.

Despite anxious people being caricatured by their physical appearance, explaining the link between attitude and appearance takes some lateral thinking. Kagan suggests that a single area of the foetus called the neural crest could be the key. It develops into ganglia of the sympathetic nervous system, and the adrenal medulla, which produces noradrenaline. Its cells also migrate into facial bone and into cells called melanocytes which produce the body's pigment. In this way, determinants of face shape, colouring, anxiety and heat generation are closely linked at an early stage of human development. Perhaps, says Kagan, changes in genes monitoring the chemistry of the neural crest in those very early northern Europeans contributed to the development of distinct physical features.

Some support for this hypothesis comes from animals. Russian geneticist D. K. Belyaev was living on a Siberian farm in the 1950s where silver foxes were bred for their fur. He selected only the tame, less fearful animals and interbred them to make life easier for the breeders. It took eighteen generations for all of the offspring to become tame. But Belyaev noted that as the litters became less fearful, levels of steroids produced by the animals' adrenal glands also fell. More astonishingly, dark-coloured melanin spots started to appear on the tame animals' coats. As the animals became more tame and less anxious, they also became darker. This suggests that the same genes regulate fear, colouring and steroid levels. Detroit neurophysiologist Tom Uhde's current work with pointer dogs, described in the next chapter, found further evidence of a link between physique and temperament. He noted that nervous dogs are consistently smaller than their bolder relatives.

Kagan's scenario is compatible with evolutionists' explanations of fear described in the previous chapter, but its focus is different. The evolutionists considered primarily the risks from wild animals, storms, the dark and so on. Reactions to threats – refusing to leave home, freezing at a cliff edge or blushing at being addressed by a social superior – are assumed to have increased people's safety, and therefore their longevity and fertility, but the reactions, then as now, are only helpful within limits. An antelope needs to find a balance between running from a lion and continuing to eat, and so do people. Anyone too afraid of a cliff edge may be unable to get back to safety; anyone too afraid of social contact could miss opportunities for satisfying relationships or promotion in the social hierarchy; anyone unable to leave the home base at all may become a liability to the group. Most evolutionists believe that fear is transmitted through our genes but, as discussed in the previous chapter, it is possible that such reactions could be learned culturally from our environment, our parents and other influential adults.

Kagan describes a population in extreme environmental conditions that would strengthen the forces of natural selection. Being able to withstand the cold would clearly be a good thing. The climate changed the gene pool so that people became warmer and may have, incidentally, become more fearful. Fearful people are more likely to freeze at cliff edges or faint at the sight of blood but, according to Kagan, a change in the gene pool tens of thousands of years ago could have been prompted by climate, not danger. The climate favoured people with higher body temperatures, and a nervous or fearful temperament was simply a side issue.

Body temperature could not be passed on culturally and Kagan's hypothesis rests on genetic transmission of a fearful temperament. He suspects that within the global population, there are 'hot spots' of races, like the northern Europeans, whose ancestors were more than usually fearful as a result of the brutally cold climate long ago. He is not suggesting that the whole of northern Europe is a bag of nerves, rather that a fearful temperament may be more common among people descended from northern Europeans than in those whose ancestors lived further south.

Of course, being fearful does not necessarily mean having a phobia. But being naturally shy and inhibited may be a powerful starting point for developing phobias, especially social phobia.

## Three Generations

Sitting on a stationary tube train and watching it fill with smoke would be bad enough for anyone, but for someone who dreads any form of public transport, it could be unendurable. Ian, who has social phobia, found himself in just that situation. He looked around the carriage and the reaction from the other passengers was muted; a few nervous remarks and concerned looks; nobody broke down in tears or showed obvious distress.

Ian got by and afterwards felt it was a personal triumph. 'I coped with the situation. I didn't really have any alternative, but I didn't feel as bad as I thought I would have done.'

Ian hates travelling by train. Sometimes he manages, sometimes not. His stomach starts churning, he feels ill, he has to get off. Crowded trains can be impossible – and worse, strangely, than a carriage filling with smoke. 'I knew that if I had reacted in another way then, nobody would have been surprised. There was less danger of losing face or appearing weird because irrational behaviour in that situation would have been more acceptable.'

Loss of face is everything for someone with social phobia. Ian fears situations he cannot easily leave, confined or crowded spaces, flying and public transport. He avoids conferences and public speaking. 'It is all part of the same thing,' he says. He is now fifty-five and has had the same fears for as long as he can remember, certainly since school. Exams were a nightmare.

He sought help only once, a few years ago, when a change at work meant he was obliged to give presentations. A clinical psychologist tried to teach him how to relax in stressful circumstances, but although Ian went regularly for eighteen months he is not convinced the sessions made a difference: 'It was easy enough to relax at home or at the hospital, but when a situation arose at work I didn't find it easy to think of myself lying on a beach with the waves lapping on the sand. I don't think it really helped.'

Over the years Ian has learnt to work round his phobia. He trained as an accountant and now works as an investment manager, so that most of his work involves seeing others individually or in small groups. Alcohol has helped him through social gatherings, especially when he was younger, but he would never think of going to football matches or rock concerts. He is married with two grown-up daughters. 'I don't have any deep resentment that I haven't achieved what I might otherwise have done,' he says.

The same cannot be said for his daughter Nicola. She feels that her social phobia has interfered at all the important times of her life. It was first apparent when she was nine years old and had to be excluded from school assembly because standing among the other children made her feel ill. She got worse over the years and by her early teens was having regular panic attacks. She would start shaking violently, feel sick and be very aware of her heart rate. Sweating and unable to breathe properly, she would become desperate to escape the situation.

Like her father, Nicola found the exam room torture and left school at sixteen with two GCSEs rather than the anticipated ten. This ruled out college and instead she went straight to work. She met and married her husband and became pregnant soon afterwards, which brought immense relief. 'I was very well in that nine months, possibly because I had something to focus on other than the phobia. I had a viable excuse and I knew that people wouldn't think I was weird, because I was pregnant.' It echoes her father's feelings about the smoke in the carriage; having something to blame other than the phobia can be a tremendous relief.

Nicola's marriage did not last and she now believes she went into it for all the wrong reasons. 'It was an escape route. It meant I did not have to go out anymore looking for a boyfriend.' She was then left to bring up her daughter, Catherine, on her own. Her phobia made outings with a young child difficult, if not impossible. Her mother would sometimes come and take them both out, but on her own Nicola found even a trip to the nearby park too much. She avoided toddler groups and found children's parties hard as she had to chat to parents she did not know.

As time went on, Nicola noticed that Catherine was becoming awkward. The little girl obviously did not like crowded rooms and found socialising difficult. Nicola worried that Catherine was following the same path as she had done, and set about changing things.

She has since received a combination of counselling and drugs and has turned her life around. Still only twenty-nine, she attended an access course which has allowed her to go on to study psychology at university. She has a new relationship and is now making sure that Catherine misses no opportunities. They go out to the cinema together, they visit relatives miles away, they have even been to EuroDisney, via the Channel tunnel – still difficult for Nicola, but wonderful for Catherine.

Anxiety-related and other disorders run through Nicola's family. Her sister has irritable bowel syndrome, and her father's sister had anorexia nervosa as a young woman. Her father's mother has told her that the trait for anxiety may go back further. One of her greatest hopes is that Catherine, who is five, is displaying normal shyness, rather than the beginning of social phobia. 'I don't want her growing up as I did. The likelihood is that she will copy me, but I don't want her to go through all this. It has spurred me on to change things and I am just hoping it is not too late for her.'

## A Family Story

Nicola's family is not alone. Epidemiologists say that if you have a relative with agoraphobia, social phobia or a specific phobia, your own risk of having the same problem doubles or triples. Results vary from study to study, but having a first-degree relative with panic disorder, closely linked with agoraphobia, pushes up your own risk from around 2 per cent in the general population, to anywhere between 10 and 20 per cent. Phobias run in families.

These figures hint that genes may be involved, but they are far from conclusive. Strong family traits can be formed entirely by the home environment. Money, religion and language run in families; none is determined by our genes. Our inclinations towards work, sport, food and drink are shaped by the prevailing attitudes at home; we are likely to have the same kind of education and social class as our siblings. The family is powerfully influential.

Scientists interested in the genetics of fear need to tease out the environmental influence – difficult enough even for apparently straightforward traits like height and weight. Weight is primarily determined by environment, and can be controlled by a combination of diet and exercise. Height is more genetically predetermined. But neither is clear-cut. Genes influence weight and environment is involved in height. Certain genes can increase the likelihood of storing fat and could be responsible for the tendency to become overweight, as with the Pima Indians mentioned in the previous chapter. Changes in diet or other environmental factors are responsible for successive increases in height over generations. The average adult height in most industrialised nations is increasing, most spectacularly in Japan, where the generation gap can be measured with a ruler. Japanese men are on average four inches taller than men were in 1947, presumably because of changes in diet. The genes determining height cannot have altered in a few decades.

In the same way as some families are tall, and some tend to be overweight, others may be prone to high anxiety. Height and weight can be easily measured, but anxiety is harder to pin down. Where does normal concern end and clinical anxiety begin? How do we compare one description with another? How do we grade fear? And where do we draw the line between fear and a phobia? We have to be able to measure anxiety reliably before we can say a family has more of it than expected.

The relationship between fear and panic is hotly debated. One side claims the genetics of each are relatively independent; the other – mainly American – side says that agoraphobia is so closely related to panic disorder that the same gene controls both. Whichever is true, searching for genes

for panic disorder has the practical advantage that it can be
reliably defined and diagnosed, and scientists agree that it
has an underlying biological cause.

Reliable diagnoses are crucial for scientists trying to follow
a trait within a family. Unless scientists can be sure who has
the disorder and who does not, they cannot hope to find the
gene causing it. A salutary tale involved researchers from
the University of Miami looking for the gene for manic
depression. They were studying a large Amish family blighted
by the condition. The scientists found that family members
with the disorder appeared to have distinctive markers on
chromosome 11, and they thought they were homing in on
the gene. However, subsequent analysis by other groups failed
to confirm their finding and they were forced to retract. They
found later that a couple of misdiagnoses may have been
sufficient to explain the mistake. If manic depression had
been diagnosed in just two other family members, it would
have eliminated the original association with the marker on
chromosome 11.

Similar problems could arise with panic disorder. The
*Diagnostic and Statistical Manual of Mental Disorders* lists
strict criteria to keep assessments relatively consistent, but
they still sound rather arbitrary. Panic attacks, for example,
a part of panic disorder, require people to experience at least
four of thirteen criteria. Symptoms, which include palpitations, chest pain, chills or hot flushes, dizziness and a fear
of dying, must reach a peak within ten minutes. A man who
suddenly felt so intensely uncomfortable that he had to sit
down is not diagnosed as having a panic attack if he cannot
be any more specific. Nor is a woman whose symptoms took
fifteen minutes to peak. Two physicians might make different
diagnoses on the same patient, and for a good reason. The

DSM itself is changeable, updated every few years with definitions of syndromes adapted as new knowledge and understanding – and sometimes even politics – are taken into account.

Researchers wanting to study a mental disorder must choose their definition carefully. A hairline difference in researchers' definitions or methods can become a chasm when they compare results. If one study conducts interviews face to face and others use telephones or questionnaires, their figures will differ. Interviewees might be more likely to remember incidents with someone prompting them.

Consequently, the available research yields a perplexing variety of figures. An American professor of psychiatric epidemiology, Myrna Weissman, set out to collate the results of different studies on panic disorder. She ruthlessly excluded studies that had not followed strict criteria and found that panic disorder affects between 1.2 and 2.4 per cent of the population at large. Having a close relative with panic disorder pushes up the risk dramatically. In one study, 8 per cent of people with an affected relative developed panic disorder themselves; in another, it was 20 per cent. The figures differ but they do emphasise the same definite trend: relatives of someone with panic disorder have an increased risk of running into the same problem themselves. This suggests that genes may be involved.

Two Australian researchers, Carmen Moran and Gavin Andrews, interviewed sixty people with agoraphobia and found that one in eight had close relatives with a history of agoraphobia. This is two or three times the normal risk, and similar to Weissman's findings. But agoraphobia did not occur repeatedly down the generations. People with the disorder had either a parent or a sibling affected, not both, and only two of the study group could describe agoraphobia in three successive generations of their family. This argues against genetic transmission of agoraphobia and suggests the family environment might be more important. Then, as if just to confound this conclusion, the researchers uncovered the case of a woman who met her biological mother for the first time nine years after developing agoraphobia. She found that her natural mother, aunt and half-brother had all experienced agoraphobia, whereas her adoptive family showed no signs of it at all. This case, in contrast to their general finding, supports the idea of genes for agoraphobia.

Other types of phobias pose similar problems for scientists trying to understand how they develop. Blood and injury phobias are a particular type of specific phobia which are more strongly rooted in the family than most. A small Swedish study found that two-thirds of people with blood phobia have relatives with the same reaction. The figure rises to 70 per cent among those who faint at the sight of blood. Injection phobias are more like other specific phobias, and only a third of those afraid of needles have relatives with the same fear. But again this figure rises – almost doubles – in people who faint. It seems that fainting makes the family connection stronger.

As mentioned in the previous chapter, fainting at the sight of blood or a needle is almost unique among phobias. Spider phobics might think they are going to pass out in fear but they almost never do. Their thumping heart and racing pulse sends blood rushing around the body at top speed which primes them to run away or stand and fight. Soaring blood pressure stops them losing consciousness. Blood or injury phobics, however, often find their blood pressure plummeting at the sight of blood or a doctor's needle and they sometimes faint in fear. This distinct reaction hints that an altered physiology is being passed on from one generation to the next and is another clue that genetics may be involved.

## Bundles and Blankets

Researcher Abby Fyer and her colleagues set up an ongoing family study of anxiety disorders in New York. Patients of European descent who had at least one close relative were recruited through the media and by word of mouth. All were interviewed directly.

They found that specific phobias of animals, crowds, heights and so on, also cluster in families. Close relatives of those who had been treated for a specific phobia had themselves three times the normal risk for specific phobia. Relatives' phobias tended to be of the same general type, but of different objects. Where a parent feared dogs, the child hated cats; or if a girl disliked the dark, her brother feared heights.

Specific phobias were distinct from other mental disorders and not linked with anxiety or depression. Being phobic of dogs, for example, does not make you any more likely to have agoraphobia or social phobia. Fyer believes that specific phobias are transmitted through families as a discrete bundle, quite independent of other mental disorders.

In another study, Fyer found that relatives of someone with social phobia were at increased risk of having social phobia but were not especially vulnerable to other anxiety disorders. Social phobia, like simple phobias, appears to be distinct from other anxiety disorders.

Her work suggests that different phobias can be regarded as separate entities. Specific phobias are not related to social

phobia, and neither are associated with other anxiety disorders. These phobias travel as if in carefully wrapped bundles
and receiving one for a simple phobia does not make you
any more likely to get social phobia. Likewise, a social phobia
bundle only increases your chances of feeling anxious in
social settings. The bundles are discrete and symptoms do
not leak out.

According to Fyer, a girl who is scared of spiders is more
than usually likely to have brothers and sisters who hate
snakes or birds. But an adult who hates spiders is unlikely
to have children who hate going to parties. Parents afraid of
heights are no more likely than anyone else to have agoraphobic relatives. Different types of phobias, be they of animals, situations or social events, are different right down
to their biological and psychiatric roots. Each has its own
contributing genes; each may be triggered by a separate type
of event or environment.

However, other research groups do not agree. They believe
conditions overlap generously. If people inherit anything,
they inherit blanket neuroticism, a general tendency to be
anxious. The phobia they are most likely to develop then
depends on their family environment. A man with social
phobia may be unable to write his name in public, and have
to carry cash with him everywhere as he cannot sign a cheque
at the supermarket till. If his children have a tendency to be
anxious, the chances are that they will also fear public scrutiny. Children with an identical biology brought up by a
mother or father with agoraphobia, would see the most
influential figure in their life afraid of leaving home, or going
into crowded streets. This role model is likely to channel
their fears in the same direction and make them cautious
about going out alone.

Both camps agree that genetics *and* environment contribute to the development of phobias. Fyer's view of a series
of discrete disorders would be backed by the discovery of a
series of different genes, each coding for a different phobia.
The other 'blanket approach' view predicts that certain genes
increase the likelihood of a range of mental disorders. Both
allow for environmental influence.

## Double Trouble

The classic technique for sifting out genetic influences from
the environment is to study twins. Identical twins develop
from a single fertilised egg which splits down the middle
early in pregnancy. They are, therefore, genetically identical.
Their biology, down to cell level, is as similar as their appearance. Non-identical twins develop when two of the mother's
eggs are fertilised at the same time. Like brothers and sisters,
they share half their genetic material, can be the same or
opposite sexes and may or may not look similar.

If an identical twin has a condition largely caused by genes,
such as cystic fibrosis, the other twin has identical genes and
is at a very high risk. The non-identical twin of a sufferer,
however, may be protected by differences in his or her genes.
When one child is diagnosed as having cystic fibrosis, a non-
identical twin – like any other brother or sister – has a much
higher risk than an unrelated passer-by, but it is still lower
than for the identical twin whose genes would *certainly* put
them at high risk of becoming ill.

Svenn Torgersen at Oslo University used the idea that if
a condition is caused by genes, identical twins should be at
more risk of developing the same disorder than non-identical

twins. He interviewed same-sex twins of people who had been hospitalised for an anxiety disorder. Identical twins were twice as likely to have an anxiety disorder themselves as the non-identical twins. This suggests that something in the genetic make-up of the identical twins is putting them at added risk. It is a strong hint that genetics are important in anxiety.

However, the results were even more striking when they were broken down into separate disorders. Identical twins had a risk of developing panic disorders, including agora-phobia with panic attacks, five times higher than non-identical twins. Their genes were apparently pumping up their risk. Ten years on, these researchers carried out a similar study and confirmed the importance of genes in panic dis-order. However, rates of social and simple phobias were simi-lar in both types of twins, which suggests that the primary influence is environmental, not genetic.

Meanwhile, Andrews in Sydney took advantage of Aus-tralia's enormous twin registry. More than 2,900 pairs of same-sex identical and non-identical twins completed and returned questionnaires. The researchers found that people with a neurotic personality were likely to suffer from anxiety and depression and they concluded that neuroticism, anxiety and depression are influenced by the same genes.

This study found that when both twins had an anxiety disorder, they often had quite different symptoms. One might complain of feeling breathless, while the other had a pounding heart. This underlines the point that genes are not the only cause of anxiety disorders. The genes might contribute to the general anxiety trait but differences in the twins' experiences mean that the anxiety manifests itself in different ways. Some highly idiosyncratic, individual experience, perhaps at a critical and sensitive time of development, might determine which symptoms eventually develop.

Researchers went on to look at six common anxiety or depressive disorders, including social phobia, obsessive-compulsive disorder and agoraphobia with panic. They interviewed the twins directly and found that identical and non-identical twins had similar rates of both developing these illnesses. There was little evidence that genes played a significant part in shaping the precise disorder from which the twins suffered.

The Australian work backs the idea of a blanket neurotic trait, which can spill out and translate into any one of a handful of anxious illnesses. The general neurosis might lead to either social phobia, agoraphobia or obsessive-compulsive disorder. It again contradicts the idea of social phobia as a discrete bundle, unrelated to other anxiety disorders. The finding that genes have only a modest impact on agoraphobia contrasts also with the Norwegian work, which showed genes pushing up the risk of panic and agoraphobia fivefold.

One possible explanation for these confusing and contradictory results is the difference in the study populations. The American work was done at a specialist clinic. The Australian subjects were taken from the general population, but in the Norwegian study, at least one of the twins had been hospitalised. It was later criticised for just this reason since the vast majority of people with anxiety disorders never receive any medical help at all. People who are hospitalised for anxiety are at the extreme end of the spectrum and may not be representative of those with the same phobias who have never sought help. But if genes are most important at the severe end of the spectrum, it could explain why they seemed to be so much more important in Torgersen's sample than in Andrews's.

Together, the studies confirm that families can foster anxiety disorders and neurosis. But the conflicting findings blur the overall picture, and give few insights into exactly how important genes are in the different phobias.

## Computer Synthesis

In a fresh attempt to make some sense of the contradictions, Kenneth Kendler, psychiatrist at the Medical College of Virginia, took a new approach to quantify the contributions made by environment and genetics to each type of phobia. Researchers interviewed 2,163 female twins, and assessed them for major depression and four types of phobia: agoraphobia, social phobia, situational phobia and simple phobias. A complex computer model analysed the data.

The computer started with four basic hypotheses for the causes of phobias. First, genes could have a blanket effect and predispose people to all types of phobia. This echoes the neuroticism idea, and suggests that general 'phobia' genes make you prone to anxiety and phobias. Second, a specific gene or genes could be responsible for each type of phobia. This reflects the idea of phobias as discrete bundles in which the genes that contribute to a fear of public speaking don't make you any more afraid of insects. Third is the blanket environment effect in which a less than satisfactory childhood, for example, increases vulnerability to all phobias. Fourth is the specific environment in which a particular event precipitates the development of a particular phobia.

The researchers fed in the data from all the interviews,

and the computer calculated which hypothesis gave the best fit for each phobia. They found different hypotheses hold true for different types of phobia. A one-off traumatic event was often paramount in the development of specific phobias, including animal phobias and situational phobias. A middle-aged woman, for example, said she would not go into water if she could not see the bottom. It seemed a peculiar and unusual fear, but when asked why, she said that as a girl she and her friends had often gone swimming in a river close to home. One afternoon she jumped into the river and landed on a corpse – a harrowing incident, and one which makes her subsequent phobia quite understandable.

Those with agoraphobia could sometimes describe a first panic attack very clearly, but the trigger for the attack was often difficult to pin down. The specific, one-off event was not particularly important for agoraphobia. The 'blanket' environment, perhaps a troubled childhood lacking in stimulation, was more important.

Social phobia appeared to lie somewhere in between. Occasionally, someone with social phobia could recite a story of some terrible gaff, such as urinating slightly while talking to a group of people, and ever afterwards being immensely afraid of public speaking. But many said they had always felt nervous and lacking in confidence in certain situations.

The models suggested that there are general genes that make people likely to suffer from some psychiatric disorder, be it a phobia or perhaps depression. These general genes influenced all phobias slightly. However, there are also genes which have an additional and specific effect only on agoraphobia.

The influences on agoraphobia and specific phobias were at opposite ends of the spectrum. Specific genes were the

most important influence on agoraphobia but a long-term
unsatisfactory environment also played a part. Genes were
relatively unimportant for specific phobias; a one-off traumatic event was the primary cause.

Other features of agoraphobia and specific phobia contrast
in a similar way. Agoraphobia typically starts late, usually in
the twenties. It is likely to be associated with other disorders,
such as depression, and cause major disruption to life. Specific phobias tend to start in childhood, are not often associated with other problems and – though they can have serious
consequences – do not usually dominate people's lives. It is
as if phobias lie along a continuum, with agoraphobia at the
more severe end, specific phobias at the less severe end and
social phobias somewhere in between.

This was the general finding from thousands of interviews.
However, an individual's phobia, of whatever type, can still
have a contribution from genes or environment. According
to Kendler, those who develop a severe phobia after a relatively mild event probably have a higher genetic component.
Compare the woman who leapt on to the dead body in the
river with another man afraid of water. He was learning to
swim as a boy when a group of bigger kids jumped in right
beside him, splashing him and making him gulp down some
water. An unpleasant experience, but not a rare one, and
Kendler says that his phobia is more likely to have a genetic
component.

A woman who had driven confidently for most of her
adult life was suddenly unable to get into a car. She could
not drive, she would not be driven and she could barely walk
past a parked car without feeling shaky. It started after she
was involved in a serious accident. She was trapped in the
wreckage of a car for forty-five minutes while people ran

around shouting that the car was about to explode. Her resulting phobia does not seem unreasonable. The exceptional one-off incident is probably sufficient explanation for her fears. Like the woman afraid of murky pools, she probably had a low genetic vulnerability to phobias and was unlucky enough to have a terrifying experience.

One initially surprising finding was that fear of tunnels, bridges, the dark and other situations were more like specific phobias than agoraphobia. Situational phobias like these have often been considered mild forms of, or undiagnosed, agoraphobia. Kendler believes it may be important that those with situational phobias, like those with animal phobias, can often describe very precisely the conditions that they fear. They may fear iron bridges over water and be unconcerned by stone bridges crossing railway lines. Or they may be afraid of the dark only when there is a strong wind blowing from the east. A tight relationship with the phobic stimulus may keep the phobia discrete, or specific, and therefore more manageable. Agoraphobia involves a more general fear, in which crowded streets or empty streets, bright lights, supermarkets and a host of other situations cause terror.

The study's conclusions are strangely diplomatic. The finding that some genes make us vulnerable to any phobia backs the idea of an inherited general tendency to anxiety which can translate into many different disorders. On the other hand, the study suggested that other genes specifically increase the risk of agoraphobia, which backs Fyer's view of different genes for different phobias, at least as far as agoraphobia is concerned.

Taken together, the family studies suggest that the genetic effect is at its strongest for the development of panic disorder and agoraphobia. Agoraphobia is also the most severe phobia

and might intuitively be assumed to have the most profound
biological, or genetic, input. The next step is to identify the
gene or genes involved and agoraphobia is the most obvious
place to start looking.

## The Hare and the Tortoise

Genes for physical disorders are being isolated with astonish-
ing frequency. A new discovery seems to be announced every
few weeks. But tracking down the genes behind psychiatric
conditions has been a long and difficult task. The genes
responsible for many of these disorders have turned out to
be intricately and confusingly packaged.

Finding the gene for Huntington's disease was a rare success. People with Huntington's develop involuntary muscular
movement, the so-called chorea, and progressive mental
deterioration. The symptoms develop in people from their
mid thirties, often after they have had children and passed
on the gene. The discovery was possible because, genetically,
Huntington's is a straightforward condition, caused by a
single dominant gene. It is now possible for those at risk to
be tested. The information could help them decide whether
to have children, though any genetic test is an ethical minefield. If a teenager gets a positive result, their parent, who
may not have wanted to take the test, will know for certain
that they also have the gene.

Huntington's disease was first described in 1872 by an
American doctor, George Huntington. As a boy, driving with
his father on Long Island, he saw two women, mother and
daughter, both tall and desperately thin. They were bowing,
twisting and grimacing and made such a spectacle that the

memory not only stayed with him, but propelled him into this field of medical research. He visited 'many a home where the bearers of the gene waited with stern Calvinistic stoicism for the dreadful fate that Providence had meted out to them'.

Huntington's disease is so dramatic that it could be reliably diagnosed long before modern techniques of molecular biology became available. In addition, early geneticists were able to draw up family trees and trace the transmission of the gene from generation to generation. But the genetics behind the main causes of premature death in Western society are more complex. In heart disease, obesity and cancer, genes and environment interact to produce illness. Many who have the genes do not get the disease and it has proved difficult to unearth the genes responsible. Furthermore, the genetic contribution to these conditions does not come from a single gene. Rather, a collection of genes spread out on different chromosomes all contribute to eventual illness. Scientists are starting to come up with answers, helped by advances in laboratory techniques, but progress in our understanding of the genetics behind these conditions remains patchy.

Psychiatric disorders present similar difficulties as the relevant genes are also likely to come from many different chromosomes. They, too, are the result of a mish-mash of interactions between genes and environment. Where one twin has panic disorder, their identical twin has only a 30 per cent chance of developing anxiety disorder with panic attacks. Genes alone are not sufficient to cause and explain the problem.

A further difficulty is our lack of understanding of the mechanisms behind psychiatric illnesses. We know the basic biology of physical disorders: in heart disease, atherosclerosis, or the furring-up of arteries, high cholesterol levels, platelets and clotting complexes in the blood all play a role; in diabetes, the importance of the hormone insulin is understood; the processes that turn cancer cells on and off have been studied. This sort of knowledge gives geneticists a starting point when they come to look for the gene.

Our theories on the systems in the brain involved in psychiatric disorders are still just theories. The human brain is extraordinarily complex and our proven knowledge is slight. Scientists do not have a definitive starting point for their search.

The inconsistent diagnosis of psychiatric disorders complicates things still further. As discussed earlier, changing the criteria used for diagnoses even slightly will skew epidemiological studies. In genetics, the difficulties are even more acute, because scientists trying to identify genes tend to be studying fewer people.

Another problem is that many different genetic flaws may lead to the same mental illness. The human central nervous system probably has a limited number of ways of reacting to disorder within the brain. Tens or even hundreds of different underlying glitches might give the same outward appearance. It's comparable to dropsy in the mid nineteenth century, then assumed to be a single condition with a single cause. In fact, oedema, as it is now known, is simply an inappropriate build-up of fluid. It is associated with several serious medical conditions including heart failure, liver disease and malnutrition, but can also be caused by simple gravity. Anyone who sits on an edge and swings their legs for long enough is likely to get swelling round the ankles. Oedema is only a sign of some underlying cause.

Likewise, mental disorders such as depression or anxiety might be just the outward symptoms of many unrelated internal disturbances, with any number of different sets of genes contributing to them. This has already been suggested for schizophrenia. In 1988, a scientific journal carried two reports. One, from a London group, described an association between schizophrenia and markers on chromosome 5. Another from Yale University described the lack of any such link.

A possible explanation for the contradiction is that the groups looked at different families. It could be that the schizophrenia in the family studied by the London group *was* linked with a genetic defect somewhere on chromosome 5. The American family may simply have a different genetic defect leading to the same disease. As it turns out, the London group's finding was not confirmed elsewhere, and it seems that the link they found was just an unfortunate coincidence. But the idea that many internal routes lead to the same external behaviour may turn out to be true for schizophrenia, and indeed, for panic disorder.

Imagine that three sets of genes, A, B and C, act together to produce panic disorder, but only two of the three sets are sufficient to put someone at risk. They could have A&B, B&C or A&C. Geneticists start their search for the gene by finding a family blighted by panic disorder. The scientists scour the family's genetic material and come across area A, which appears to be involved. They begin to think they have found the genes. In order to back up their work, they take another family with a strong history of panic disorder. But this family have the B&C genes. When the researchers look for oddities in area A, they find nothing and have to assume that their original finding was a statistical fluke, or that somehow they were mistaken.

Were there only three areas to be found, such problems

would be overcome relatively easily. But nobody knows how many will turn out to be involved in panic disorder and, not surprisingly perhaps, initial studies searching for genes associated with the disorder have been disappointing.

A group led by Raymond Crowe in Iowa has now scoured 70 per cent of the human genome for genes associated with panic disorder and found little to excite them. Gene hunters first divide human genetic material down into small stretches and then look within each stretch for their target gene or genes. LOD scores are a measure of how likely it is that these genes lie within a particular stretch. A LOD score of 3.0 is strong enough for researchers to sit up and take note. The Iowa group's first trawl for panic disorder genes uncovered only a few areas with LOD scores over 1.0, and none as high as 2.0. At best, this means the areas deserve further study.

That does not mean the genes are not there. The family and twin studies have convinced researchers that genes are important. But they are going to be difficult to find. The next step for the Iowa group is to look through the genetic material again, this time with a finer map, breaking the material down into smaller chunks. It will be a long haul. They believe it will take them three years to produce a map with double the resolution of the first, and they are unwilling even to guess when they will start getting results.

Technological advances in the meantime will help as this painstaking work plods forward, tortoise-like. But it is not the only way to find the genes and a more hare-like approach could hit lucky and get there first. This other approach relies on finding 'anxiety' or 'emotionality' genes in animals and then looking directly at the human genome, to see whether genes in the same areas might also be associated with anxiety in humans.

This technique was used successfully in diabetes. Until about eight years ago, the prevailing wisdom was that diabetes was associated with up to a hundred genes, each making a tiny contribution. But John Todd in Oxford started looking at mouse models for diabetes and was able to demonstrate that a handful of genes only were responsible for most of the effect. There was a tailing-off effect, with many more genes having a smaller effect, but five or six key genes held by far the most influence. Todd later showed that the same was true in humans. At a stroke, this work pinned diabetes down. Suddenly, we knew that only a few genes were the root cause of most of the problems, and it became possible to look at diabetes in a new light: as a genetically transmitted disease.

Psychiatric disorders like panic disorder are also thought to be influenced by many different genes. A colleague of Todd's at Oxford, Jonathan Flint, wondered whether an approach similar to the one used in diabetes might give similar insights. Initial results suggest he might be right. It is difficult to study psychiatric disorders with animal models since behaviour is not easily interpreted. We cannot ask animals what they are feeling and virtually any observation is open to conflicting explanations. In a basic experiment with mice, for example, experimenters shine a light at the animals and then deliver an unpleasant but not painful electric shock to the animal's foot. Most of the animals learn that the light means the shock is coming, and with true Pavlovian conditioning they run away as soon as the light comes on. But some of the animals stay where they are. One interpretation is that the experiment is measuring animals' fear, and those that stay are not as afraid of the shock. An equally valid view could be that their feet are insensitive to pain.

To get round this, experimenters breed animals according to their performance in one experiment and then cross-test them in different circumstances. In collaboration with Flint, John DeFries at the Institute of Behavioural Genetics, Boulder, Colorado, wanted to breed anxious and non-anxious mice. He found that anxious mice in an open field defecate and cower, they do not run about. In other tests, these same anxious mice are reluctant to explore the open arms of a maze and also slow to jump across a box to avoid a shock. Non-anxious mice show the opposite traits.

De Fries kept selecting out anxious and non-anxious mice and breeding them with other animals with the same trait. After several generations, the mice bred true, and all of the offspring behaved in the same way as their parents. He then bred the anxious mice with non-anxious mice. This first generation was uniform, each mouse with one anxiety and one non-anxiety copy of the genetic material. When they were interbred again, the offspring were a mixture. Some had two copies of an 'anxiety gene', others had two 'non-anxiety' genes and others had one of each. The genes influenced the animal's behaviour and their score on the tests.

Flint and his colleagues then analysed the genetic material of the mice who had high scores in tests and were presumably carrying two 'anxiety' genes. On their first look through, they found six areas of potential interest. Further tests enabled them to say they believe there is an important gene in an area on chromosome 1.

As with diabetes, it is still true that many genes may have an impact on anxiety scores but, in mice at least, a single area on the chromosome may prove to be as important as all of the rest of the genes put together. This tantalising finding hints that the genetics of psychiatric disorders may be more manageable than previously thought. It could be a false dawn – mice are not men – but it is hopeful.

The first step will be to see whether this area is also important in rats. If it is the same in two species, it may be time to look in humans. Chromosome 1 in mice correlates for other genes with chromosome 1 in humans and the work hints that we may eventually be able to get to the answer. However the puzzle is far from solved.

Genes for panic disorder will be found in this century, either by the methodical approach or by a lucky hit, and however it happens it will create great excitement among the research community. Researchers, though, are united in their caution for what it would mean for people with phobias. Finding the gene does not automatically, and certainly not immediately, produce new treatments. The genetic mutations responsible for cystic fibrosis have been known for years, but new treatments are just starting to become available. The gene for Huntington's disease was found ten years ago, but the discovery has not yet spawned new treatments.

The discovery of panic genes would immediately prompt further research. Scientists would finally be able to find out whether those with the genes are more vulnerable to social and specific phobias, as well as to agoraphobia. This would be a big step towards understanding the overlap and the distinguishing features of these phobias.

But it will be a long, long haul before we have new treatments as a direct result of finding relevant genes. Once the genes are isolated, scientists will discover which proteins they make and set about mimicking them in the laboratory. There are many obstacles to be overcome before new drugs designed in this way become available, but it is possible that one day we will have drugs that fit into specific pockets in the brain, reducing fear with no other side effects. Current drugs have been found by chance and we often have little idea how or why they work. These new drugs might prove to be cleaner, with highly specific effects. It would be a big advance, particularly for those most severely affected by phobias.

Today's talking treatments will still remain valid once we have found the genes. Genes will never be the sole, or even the main cause of phobias, but genetic research may, over years to come, revolutionise our understanding of the biology of fear.

## CHAPTER 4

# Neurophysiology

## *Personal Magnetism*

Counting to ten is not usually a challenge for Mark George, neurologist and specialist in brain imaging. But sitting in a laboratory with a magnet moving over his head, he found himself incoherent. He started off confidently – 'One, two, three' – but as the magnet passed over the area of his brain responsible for speech (Broca's area), he was struggling – 'Or, aye, ih.' He says it was a weird sensation. He was still forming the numbers in his head and he could hear that he was getting them wrong, but as long as the magnet remained on the spot he could not correct himself. The connection between thought and action – between thinking the number and saying it – was broken. But as soon as the magnet moved away, he could immediately continue, 'seven, eight, nine'.

George's interest in a technique called Transcranial Magnetic Stimulation (TMS) started when he worked for a year at the Institute of Neurology in London. There he met John Rothwell, who was pioneering research into TMS, but looking primarily at motor systems. He had demonstrated that putting a magnet to someone's head could make their thumb twitch. Once back in Charleston, South Carolina, George pursued the work, but rather than looking at reflex responses, he was interested in knowing whether magnets could alter how the brain works and what we can do. The counting experiment was the first indication that it could.

Dramatic though the demonstration was, especially for the subject, its relevance for people with phobias is not immediately obvious. But in fact, George's work is moving quite steadily in that direction and he is excited by TMS's potential for treating anxiety disorders.

Scientists have believed for decades that the left hemisphere of the brain is responsible for happiness and other positive emotions, the right for fear, anxiety, disgust, anger and sadness. As far back as 1960, doctors anaesthetised one or other side of the brain before surgery to find out which was more important in language, so that they could avoid it while operating. As a side effect, it was noticed that when the right was temporarily out of action people felt euphoric. When the left was taken out, they felt sad.

Other observations support this. Brain-imaging studies of the clinically depressed have shown abnormal function in the prefrontal lobes, more often on the left than the right. People who have had strokes are more likely to feel depressed if the left prefrontal lobes are damaged. And those with multiple sclerosis who are depressed have more signs of damage on the left side of their brains than those with the same condition who are not depressed. Inactivity of or damage to the left side of the brain seems to block normal feelings of happiness.

Not all researchers agree that the split between the two hemispheres of the brain divide our feelings so neatly, so George and his team set out to explore the effect of magnetism on emotions in people who are not depressed. They found that a session of TMS can alter mood. When the right prefrontal lobe was stimulated with the magnetic field, people said they felt happy. They described a strange type of happiness, a driven, almost anxious high mood. When the left prefrontal lobe was under the magnet, they said they felt transiently sad. This seems to back up the earlier assumptions about the functions of left and right hemispheres. George's work suggests that the happiness/sadness effect may be specific to the prefrontal lobes rather than a general effect throughout the brain. A later experiment stimulated the whole of each side of the brain in turn. This time, the magnetic field did not induce these artificial feelings of happiness or sadness but stimulation of the right hemisphere increased volunteers' anxiety.

This is all very interesting, but can it be applied to people who feel habitually sad or depressed? Possibly. Again, this work is at an early stage, but researchers who have scanned the brains of depressed people have commonly found that a common feature of depression seems to be that the left prefrontal cortex is insufficiently active – as if the volume button on someone's happiness and positive emotions has been turned right down.

Many with depression receive a lot of help from drugs. But some have such entrenched depression that the drugs make no difference. And it was in this group that George started to try out TMS as a therapy.

Stimulation of the right prefrontal cortex left them feeling anxious and low. Stimulation of the left prefrontal cortex provided relief. After a series of daily sessions with TMS, the group as a whole felt better. For some, the improvement was dramatic. Two had a definite and steady mood improvement over a period of weeks. One had complete freedom from depression for the first time in three years.

Encouraged, a colleague of George's, Greenburg, went on to use TMS in people with obsessive-compulsive disorder. This is a deep-rooted anxiety disorder that overlaps with some phobias, commonly fear of dirt. Some agoraphobics perform extensive checks and rituals before leaving home in a way more typical of obsessive-compulsive disorder.

This time, Greenburg found that stimulation of the right prefrontal cortex could relieve obsessive-compulsive disorders. Sufferers felt better and described the same sort of happiness boost that the healthy volunteers had experienced. They also had a general improvement in their obsessions and compulsions, and again, some got a lot better. One man, for example, had a problem walking through doorways. He always went through them backwards and insisted on using his left foot first. After TMS he walked through doorways normally for the first time in years.

Work on anxiety disorders is still limited but these results were a tremendous boost and George's team is to start exploring the use of TMS in anxiety. It is an exciting time for this fascinating new tool and discoveries are coming so quickly that it may not take long before our understanding of both the tool and the brain takes a whole new step forward.

Current knowledge about how TMS works is rather sketchy. The brain and nervous systems are electrical organs so it is not surprising that they are affected by magnetic fields. In any electrically-run gadget, from a car or generator to a power plant, electricity and magnetism are constantly being converted back and forth. The same is true in the brain and it is possible to measure both the electricity and the magnetic fields it quite naturally sets up.

Normally, an electric impulse travels along a nerve until it gets to the junction between nerves, the synapse. At the end of the nerve the impulse makes the nerve fire, shooting out tiny droplets of a chemical hormone. The droplets travel across the space between nerves and set up an impulse at the next nerve. And so it continues.

The magnetic field set up by TMS is many tens of thousands of times stronger than the brain's normal magnetism. It is produced by running electric currents through specially shaped coils and can be focused to within a few millimetres, made to pass through the skull and alight on specific centres in the brain. George says that TMS sets up an artificial electric impulse in the nerve and makes it fire. When nerves are firing, there is an on-the-spot increase in blood flow and glucose metabolism. Scientists have observed both of these effects and this backs up their working model of what is going on.

Nerves need to recover each time they fire. If the first pulse of magnetism makes the nerve fire, the timing of the second pulse is crucial. If the second pulse is given after the nerve has recovered, it will make the nerve fire again and in this way TMS can increase activity in the brain. However, if the pulse comes *before* the nerve has fully recovered, the nerve will be unable to fire a second time and will have to start its recovery all over again. It is rather like trying to flush a toilet before the cistern has filled. Repeated attempts will all fail if the cistern is never allowed to fill up. In this way, TMS can also be used to decrease activity in the brain. Its action depends on the frequency of the pulses.

Drugs which increase the supply of serotonin in the brain can lift depression and ease anxiety. George speculates that TMS could act in a similar way, perhaps more selectively. It may be that TMS can stimulate certain nerves to release serotonin so that the hormone is available precisely where it is needed.

Like many scientists and clinicians, George believes the way we classify disorders like depression is so rough as to be almost meaningless. He believes that with tools like TMS, used alongside imaging, we might begin to get a clearer idea of exactly which areas of the brain are involved in different mental disorders. We use the word depression as if it is a single disorder. He believes we will eventually find that there are at least five or ten different 'flavours' of depression and each may involve different areas of the brain, or affect them in different ways.

Whether TMS will ever become a fully fledged treatment for anxiety disorders or depression is quite a different question. For a start, it is not clear whether the effects of the magnetic pulses are long term. At a trivial level, almost everything we do leaves a long-term fingerprint on the brain. If we can remember a walk along the beach from years ago, then it has changed our brain in some way. But that does not mean a magnetic pulse can fix faulty nerve circuits in the brain and cure disorders outright.

Other treatments do have lasting effects on the brain, for example, electroconvulsive therapy (ECT) for depression. This is only given for severe and deep-rooted depression where other treatments have failed. Electrodes are attached, usually just above the eyes, and such a powerful pulse of electricity applied that it causes a seizure. Doctors have little idea why this process should lift depression, but it can be immediately and dramatically successful.

The prevailing dogma rules that a seizure is necessary for ECT to be effective. George's work challenges this idea. His volunteers became substantially better and none had seizures. On the other hand, TMS did not necessarily help them long term. They improved while having the sessions, but over a period of weeks or months slumped back into their depression. So far, George believes that TMS probably has only short-term effects. Other scientists have exposed rats to stimuli thousands of times more potent than human volunteers have received. Examination of their brains after death suggests no long-term changes.

If this is true, it means that treatment with TMS will be an ongoing process requiring repeat sessions. It is tempting to imagine home-based kits, similar to the light boxes used regularly to treat seasonal affective disorder through the winter. People could perhaps slip on their personally adapted TMS helmet for a few minutes each day.

But TMS is not as innocuous as light. The strength of the magnetic fields used in this work is potentially dangerous. The area of the brain stimulated in those with deep depression is extremely close to another area that would give a full-blown seizure if it were stimulated. For the foreseeable future at least, TMS is a research tool that can only be used as a therapy under closely controlled supervision. It is, however, a powerful research tool, and set to be used in many disorders. At present the scope for TMS is limited by the power of the magnetic fields it can produce. It can only be used to explore the outer surfaces of the brain. But multicoils are producing more penetrating fields which may in future allow scientists to search deep into the brain and open up an entirely new world of possibilities for research and treatment.

Current TMS projects that have already shown promising results range from the treatment of stuttering to the characteristic twitch or tic of Tourette's syndrome. Anxiety disorders are obvious candidates for exploration and the work on obsessive-compulsive disorder is an extremely encouraging start.

Psychiatry and neurology have been viewed with suspicion by doctors in other disciplines. The brain is such a mysterious organ, our understanding of it so limited and, until recently, the treatment options psychiatrists had were so restricted that it is not difficult to see why. These imaging techniques are helping to change that. Such advances in technology have contributed to more certain diagnoses and more scientifically grounded treatments. We have come a long way in the past 130 years.

## Hearts and Minds (and Lungs)

Soldier after soldier presented himself to the doctor. Three hundred of them in all, each with palpitations and dizziness following fighting in the American Civil War. American physician J. M. DaCosta was at a loss as to explain exactly what was wrong, but he thought they had a problem with heart and circulation and described the condition as 'irritable heart' in a publication in 1871.

It was an influential diagnosis. In the First World War, soldiers with these symptoms were still being variously diagnosed as having disordered action of the heart or a disease of the heart valves. But London physician Alan Drury, working at Hampstead Military Heart Hospital in 1916, was aware of a possible psychological dimension and noted that most of his patients were nervous and unable to concentrate. Today, many would be diagnosed as having post-traumatic stress disorder, panic disorder or other mental disturbances which had been triggered by their experiences on the battlefield.

Drury, though, according to contemporary wisdom, was

confronted by a problem affecting the heart. He was percep-
tive enough to examine aspects of patients' breathing and, in
doing so, set in train lines of research still being explored.
Drury designed a series of experiments to compare the breathing patterns of his patients with those of healthy soldiers.

He found differences. For a start, men with 'irritable heart'
could not hold their breath as long. Then, when asked to
breathe in and out of a large bag for as long as possible, they
could tolerate only half the concentration of carbon dioxide.
They had lower levels of carbon dioxide in exhaled air after
resting. On exercising, these carbon dioxide levels went down
initially in all soldiers and then rose once the exercise was
finished. However, mild exercise in the men with 'irritable
heart' was enough to produce the same response as very
taxing exercise among the healthy.

The meaning of these findings was not obvious and Drury
did not comment when he wrote them up for the medical
journal *Heart*. Another English doctor, T. Lewis, was working
at the same time and he described 'effort syndrome' among
soldiers with similar symptoms, whose problems were triggered by physical exertion.

More than eighty years on, the connection between panic
and breathing is still being debated. Many panic symptoms
– such as hyperventilation, light-headedness, dizziness or
choking – look like shortness of breath. Some researchers
say these symptoms may be fundamental to the underlying
disorder. They believe that the way people breathe even when
they are resting may hint at their chances of having a panic
attack or developing panic disorder. Other scientists go
further. One theory suggests that a panic attack has less to do
with fear than a false belief that the body is being suffocated.

Drury's work is part of the long-standing search to

pinpoint differences in the biology of those with panic disorder and those without. The list of these differences has
expanded to a bewildering degree over the years. One study
suggested that people with panic disorder have a distinct
reaction to the gentle physical challenge of standing up. Other
researchers have found that those at risk of panic can be
identified by their response to sleep deprivation, to a hard
exercise session – even to a cup of coffee. An array of unrelated chemicals given in the laboratory consistently provoke
panic in those with panic disorder and not in those without.

Some of these triggers may be interrelated. Caffeine in
coffee could cause insomnia and operate via sleep deprivation, for example. Improved understanding of such triggers
could help reveal mechanisms for panic and clarify what
panic is, what causes it and how we might successfully
intervene.

Some may be directly related to the natural process of
panic. One of the most reliable is lactate, one of the body's
waste products, which normally builds up in the muscles
during heavy exercise. As far back as the 1940s, doctors
noticed that anxious patients produce more lactate on exercise than their non-anxious counterparts. Twenty years later,
Pitts and McClure suspected that anxiety might be caused
by rising lactate levels. They tested their idea by infusing
sodium lactate into patients. Most of those with panic disorder panicked, but those without did not. It was a landmark
discovery. Here was a chemical that tipped people prone to
panic over the edge, but seemed to leave others unmoved.
It meant the biological contribution to the disorder could
no longer be disputed since panic could be reliably produced
by infusing a chemical.

The experiment has been repeated by groups all over the

world, and the original finding backed up many times – those with panic disorder react to lactate and those with other anxious or depressive disorders do not.

The finding heralded a new era in research into the nature of panic. Attacks could be produced almost on demand in the laboratory, which meant that panic could be planned and studied in minute detail. This was tremendously exciting, and looked set to answer queries on the nature of panic. But thirty years on, we still do not know exactly why the lactate produces this response. Explanations vary and include changes in calcium levels or the acid-alkaline balance in body fluids. It could work on serotonin levels in the brain, act via respiratory centres or increase general arousal levels. All the theories have some experimental evidence backing them up, but none has been universally accepted.

But work goes on. A group at Columbia University in New York is convinced of the importance of breathing to panic attacks. They believe that sodium bicarbonate, produced in the body when lactate is broken down, makes blood fluids alkaline and stimulates respiratory centres in the brain, leading to hyperventilation and panic. Carbon dioxide, another breakdown product, might work independently elsewhere in the brain to increase arousal levels and activate another pathway to panic.

The Columbia group is using ambulatory monitors to explore the link between breathing and panic, recording heart rate and aspects of breathing twenty-four hours a day. Early results show that people with panic disorder have chaotic and irregular breathing, sighing, gulping and yawning more than others. We all yawn and sigh, but over a prolonged period most have a certain rhythm and regularity in their breathing. Panic patients do not.

The group is intending to monitor between twenty and thirty patients a year in this way, and take note of what happens to their breathing and their heart when they are in difficult situations and feeling anxious. One of this study's great assets is that it will examine people as they go about their lives. It could be that panic produced in the laboratory, however genuine it seems, is fundamentally different from everyday reactions in supermarkets, buses or busy streets. People may appear to have all the same symptoms, but the artificial trigger for the attack might in itself change some underlying features. It could, for example, invoke different pathways or chemical messengers in the brain.

Ambulatory monitoring should shed light on differences between spontaneous and laboratory panic. It will also help researchers examine any special features of panickers' breathing as they go about everyday activities. If someone breathes chaotically, and also has frequent panic attacks, one does not necessarily cause the other. Chaotic breathing might contribute to panic disorder but, equally, it could develop as a result of frequent panic attacks. Alternatively, both may be caused by something else altogether.

The study may reveal why panic attacks stop. Attacks are self-limiting and, however unpleasant, relatively short-lived. The Columbia group hypothesise that hyperventilation eliminates carbon dioxide, which eventually stimulates receptors in the brain to call a halt. In this way, the panic attack could almost be seen as a natural defence mechanism.

## A Breath of Fresh Air

Don Klein has led much of the work by the Columbia group and is an acknowledged leader in the world of anxiety research. He goes further than some of his colleagues and suggests that a panic attack may occur when the brain falsely believes that the body is being suffocated. He has spoken and written widely on his 'suffocation false alarm' theory and though some believe it a victory of vision over common-sense, he tells an intriguing story.

First, and most radically, Klein says that panic has little to do with fear. Instead, he believes that panic is a biological reaction, intimately associated with breathing. Hyperventilation and feelings of suffocation are integral to panic, and have nothing to do with real fear, he says. 'The argument that panic is about fear makes no sense. The reality is ignored. Everything in psychotherapy is seen as fear and treated as misinformed fear. But it isn't fear.'

He interviewed veterans of the Second World War who were still suffering many years later. They were disturbed, certainly, they had palpitations, trembling and sweating. But they did not have problems with their breathing. At times of extreme danger, we may react with a racing heart, dilated pupils and pale, sweaty skin, all reactions designed to aid us in our fight or flight. Pale skin means that blood is being conserved for the tissues that need it most, the fast heartbeat means the tissues are being amply supplied with oxygenated blood. We do not gasp for breath.

Further, lactate and carbon dioxide are involved in respiration and are both potential agents of panic. Klein believes the physiological evidence suggests the body thinks it is being

suffocated – as if some kind of biological alarm for suffocation is going off. He cites the experience of a Turkish
torture victim. This man was subjected to many terrible
threats, including firing squads on the verge of shooting.
Not surprisingly, he later suffered from post-traumatic stress
disorder. But the assault that appeared to be responsible for
most of the enduring problems was being held with his head
under water. Attempted suffocation had severe and more
long-lasting effects than other forms of torture.

As Klein developed his theory, he wondered how he could
prove the existence of a suffocation alarm system. Every
system in the body depends on oxygen and during suffocation
about 90,000 bodily processes are being impaired. Any one
of these could sound an alarm, but Klein was convinced
there was some central alarm system for suffocation. He then
stumbled across a group of children with congenital central
hyperventilation syndrome. These children may be healthy
through the day and go to bed breathing normally. But if
they stop breathing during the night they will not react and,
unless someone else wakes them, they are likely to die. The
syndrome is fortunately rare, affecting one in 100,000 births,
and could account for some cot deaths. Children affected
were once kept alive with tracheotomy, in which an artificial
opening was made into their windpipe. Now devices monitoring the activity of nerves associated with breathing can be
implanted.

These children, according to Klein, have panic disorder
inside out. If panic disorder is an over-sensitive suffocation
alarm, these children have no alarm at all. Drugs like imipramine, which lower the threshold of the suffocation alarm,
have no effect in healthy people but stop panic in those with
panic disorder. But if children with this syndrome were to

take it, they would stop breathing altogether. Drugs which stimulate breathing may provoke panic in susceptible patients but will help these children breathe more easily.

Klein started speculating that if the pharmacology is inside out, then maybe the psychology is inside out too. Again, circumstantial evidence backs him up. Children with congenital central hyperventilation syndrome live in what most of us would find exceptionally stressful circumstances. Their lives may depend on a machine as they sleep, normally minor infections could prove fatal and they probably spend a week in hospital every year. Yet they are quite definitely less anxious than most children.

There are, of course, problems with the theory. The most obvious is the implication that whenever a human is suffocating he or she will panic. It is usually true. But hundreds die every year from carbon monoxide poisoning. A heater malfunctions and they do not react. No gulps or gasping for breath; they just fade away and do not wake up. This does not appear to fit with Klein's theory.

But the story does not end there. Evidence now suggests that carbon dioxide may be a neurotransmitter, carrying messages to parts of the brain and regulating some of its activities. Carbon dioxide may be the messenger that triggers the suffocation alarm in panic. Carbon monoxide is a close relation, consisting of a carbon atom combined with one, rather than two, oxygen atoms. Klein speculates that carbon monoxide may interact with the alarm, but disable it rather than set it off. And if carbon monoxide can switch off the alarm that triggers panic, it may itself be an anti-panic agent.

Research is now underway to test this theory. Volunteers with panic disorder will be given carbon dioxide to make them panic. Carbon monoxide will then be mixed into the air they are breathing to see whether it stops the panic attack. If it does, it will be a boost to Klein's theory and promote more work along these lines. Doctors will not, obviously, prescribe a highly dangerous gas as treatment for panic disorder, but scientists would aim to imitate its effect with safe compounds.

Along the same lines, Naomi Breslau in Detroit interviewed 5,000 people and reviewed the group one year on to identify those who had developed panic disorder. Each was paired up with another of similar age and background who had remained healthy. Breslau then looked at differences between the pair and found that one of the most important was their smoking habits. Smokers were much more likely to develop panic disorder than non-smokers.

This does not mean that smoking causes panic. But smokers often talk of the calming effect of cigarettes. Anecdotally, psychiatrists treating smokers with panic disorders have noted that if their patient stops smoking the panic often gets worse. According to Klein, someone who is nervous but who has never had a panic attack may be using the smoke as an anti-panic agent. Smokers breathe in carbon monoxide as they inhale – they could be treating themselves with carbon monoxide and warding off panic as they smoke.

One last piece of evidence to support Klein comes from an unexpected quarter. An international, multi-centre study was set up to compare the effectiveness of two drugs in treating panic disorder. Alprazolam, a benzodiazepine used primarily in anxiety disorders, was tested head-to-head with imipramine, used extensively to treat both anxiety and depression. A clear answer, as always, proved elusive, and some patients responded better to one of the drugs, some to the other. Sydney Brandon in the UK thought it might be possible to predict which patients would respond to which drug.

He maintains that patients in the study could be crudely separated according to their symptoms. One group had respiratory symptoms such as shortness of breath, choking, suffocation and a fear of dying during attacks. The others were over-aroused and described palpitations, racing heart, sweating or faintness. There were other differences between the groups. Those who had respiratory symptoms tended to have unpredictable spontaneous attacks, anywhere, any time. They did best on alprazolam. Those who were over-aroused, on the other hand, were more likely to panic in specifically feared situations and they benefited most from imipramine.

The two groups of patients may in fact have different underlying conditions, says Brandon. Panic disorder is a generalisation and only our lack of knowledge allows us to lump them together under the same heading. Those with respiratory symptoms may fit Klein's scheme, but those who are over-aroused may have a different problem altogether.

This divide is not absolute. Many patients experience both types of symptoms, and other researchers disagree with Brandon. Klein's theory may never be proven one way or the other. But the involvement of the respiratory system in panic disorder is likely to remain an enduring research question. Like so much other work in the broad field of anxiety, it will not be the whole answer for all patients. But it may be a step forward for some.

---
### *Ready for Anything*
---

In a world in which we are constantly urged to insure against everything from death, sickness and redundancy to hitting a passer-by on a golf course, few question the wisdom of being prepared. E. M. Forster in his classic novel *Howard's End* eloquently disagrees. Preparedness, he says, is a tragedy:

> With infinite effort we nerve ourselves for a crisis that never comes. The most successful career must show a waste of strength that might have removed mountains, and the most unsuccessful is not that of the man who is taken unprepared, but of him who has prepared and is never taken. On a tragedy of that kind our national morality is duly silent. It assumes that preparation against danger is in itself a good, and that men, like nations, are the better for staggering through life fully armed. The tragedy of preparedness has scarcely been handled, save by the Greeks. Life is indeed dangerous, but not in the way morality would have us believe. It is indeed unmanageable, but the essence of it is not a battle. It is unmanageable because it is a romance, and its essence is romantic beauty.

The waste of resources described here has telling echoes in anxiety. Needless worry about impending danger, disaster or embarrassment takes a terrible toll in pain, dread and missed opportunities. Phobias magnify minuscule or imagined risks and can ruin potentially wonderful events. Anyone *could* get a bad reaction from the other guests at a party, but most devote little time or energy to this remote possibility.

The social phobic, however, is alert and ready to react, constantly vigilant, examining innocent exchanges for evidence that the imagined catastrophe is about to occur. He or she devotes energy to preparing for a disaster which never happens. And as E. M. Forster put it, the tragedy is in the waste of resources.

Over-preparedness or over-arousal can be simulated in the laboratory. Experimenters have come up with a wealth of ways of putting volunteers under stress, from a painful electric shock or watching horror movies to doing arithmetic tests or puzzles against the clock. Anyone, anxious or not, reacts to the tests in ways that can be measured. Arousal increases blood pressure, heart and respiration rates. Skin becomes slightly damp with sweat, which changes its ability to conduct electricity in a way that is easily measured.

Whichever method is chosen, people with anxiety disorders take consistently longer to get used to the initial shock and give more dramatic results than those without. Non-anxious controls soon get used to the artificial stresses, they relax and their stress score falls. It has thus become accepted that having anxiety means a more violent reaction to stressful situations.

An unexpected finding from the State University of New York challenges this view. David Barlow, highly respected anxiety expert, gathered a group of people with agoraphobia, and matched each with a non-anxious control of similar age and physical condition. They each walked alone from the clinic into a busy area of town, a stressful task for someone with agoraphobia. Volunteers' heart rates were monitored as they walked, and they recorded their feelings on a tape recorder. They repeated this exercise seven times within a couple of weeks.

Barlow expected to see higher heart rates among the agora-phobics and he was not disappointed: they had a higher mean heart rate for every walk. More surprisingly, both groups showed less reaction as the experiment progressed. Non-anxious people, who reported no signs of distress, in fact had raised heart rates for the first three walks. Rates then steadily decreased. The pattern shown by the agoraphobics was similar, with a high initial mean heart rate. By the fifth attempt, the agoraphobics' heart rate had come down so far from this initial peak that it was similar to the non-anxious group's early walks.

Heart rates declined similarly in both groups and Barlow became convinced that the difference might be explained by ordinary heart rates. He was right. When the resting heart rate was subtracted from the rate during the walk, there was no difference between the two groups. It suggested that agoraphobics have a higher heart rate during a stressful test only because they have a higher heart rate before they even set out. A test designed to arouse their fears elicits no more response from them than from anyone else. It is just that they are chronically over-aroused.

Many other experiments have since backed this finding and gone on to show that volunteers made slightly anxious in the laboratory will give exactly the same response as an anxious person in their everyday, over-aroused state. Quite what causes this over-arousal is not obvious. Barlow suggests that agoraphobics may be chronically over-vigilant, worrying about how to meet life's demands. They may be specifically concerned about the tasks they are about to undertake. Or their biology might be permanently hyperactive.

## *Adrenaline*

More than ten years ago, Wendy, who works in a mill, laughed with her colleagues as the wind howled around the building. 'We kept working, joking about it. We said it was like the *Titanic*,' she says. Then the side and roof blew off the building and everyone was sent home.

Since then, Wendy has hated the wind. From October to March, she worries about the weather because winds are likely to be most severe. She never listens to the forecast, and even avoids friends on buses in case they mention a predicted storm. When the wind is bad, she cannot concentrate on anything else. 'I would hear the wind first even if somebody dropped down dead,' she says. Her fear is with her every day of her life. If she is watching a film on TV, she will be monitoring the trees outside to check exactly how windy it is. At work, in a noisy, rattling building, it is worse. 'As soon as it starts getting a bit windy, I am alert. I have got to go to the toilet. I make excuses to get out. My heart starts going fifty to the dozen, I panic. I am sweating and shaking, I start walking around, I can't keep still.'

Wendy's vigilance throughout the winter sounds like the chronic over-arousal Barlow described in agoraphobics. Her more intense feelings when the wind gets up suggest her body is preparing itself for danger. Alarms are going off inside and her adrenaline levels are surging. Scientists in the laboratory have attempted to mimic this reaction, and found that adrenaline or its relation, noradrenaline, infused into volunteers' bloodstreams, causes anxiety and sometimes provokes panic attacks.

But this is a complicated finding. Neither adrenaline nor

noradrenaline can cross the natural barrier that separates body and brain circulation. Anxiety is a feeling, an emotion, a state of mind. If these chemicals cannot get to the brain, how can they alter the way people feel? One explanation, explored further in chapter 6, is that noradrenaline acts on the body's systems to cause physical symptoms like a racing heart and fluttering stomach. These sensations are associated with anxiety and the volunteer may become worried about the physical effects. A vicious circle is set up, in which genuine concern about artificially induced symptoms exaggerates the symptoms, causing more anxiety. This leads to more prominent symptoms and yet more anxiety.

Other evidence for the involvement of noradrenaline in panic comes from its waste products. Adrenaline and noradrenaline are created and broken down rapidly so they are difficult for scientists to track. But the main by-product, MHPG, or 3-methoxy-4-hyroxyphenyletylene glycol to give its full name, hangs around much longer and can be measured more easily. When people with panic disorder were presented with a phobic object, their levels of MHPG increased, suggesting that their noradrenaline had shot up. The more fear they reported, the higher the recorded levels of MHPG and, presumably, the more noradrenaline they released.

Researchers were first prompted to study the noradrenergic system by a series of classic experiments in the mid 1970s. Researchers working with a group of stump-tailed monkeys stimulated a specific area of their brains, called the locus coeruleus. The animals reacted exactly as if they were having a human panic attack. When connections in this area were severed, the fearful behaviour stopped.

The locus coeruleus in humans and other species is situated

in the ancient centre of the brain and is associated with primitive responses. It developed early in evolution and appears to be involved in fear reactions in all species. And neurones projecting from the locus coeruleus release noradrenaline into the central nervous system.

Some drugs act directly on this system. Yohimbine, originally extracted from the bark of the rubaceae and related trees, activates the locus coeruleus to increase its firing rate. When a mixed group of people are given the drug, those with panic disorder become more anxious and have more panic attacks than those without. Clonidine acts on the same receptors as yohimbine, but has the opposite effect. It decreases the firing from the locus coeruleus and calms people who are anxious. Like yohimbine, its effect is more pronounced in those with panic disorder than those who are not anxious.

People with panic disorder seem to have an exaggerated response to compounds acting on these receptors. It might be that the receptors have become more or less sensitive to stimulation; or that other systems which normally regulate the activity of the locus coeruleus have failed in some way. Whatever the final explanation, it is likely that the locus coeruleus plays a key role in the generation of panic.

## Sleep No More

But if over-vigilance is a problem for many like Wendy, spare a thought for those who, quite literally, cannot relax at all. American researchers have recently identified a group who have panic attacks in their sleep. Just when they are feeling warm, comfortable, relaxed and unthreatened, they panic and wake up in a terrible state.

The attacks typically occur as they fall into deep sleep. They are not associated with dreams or nightmares, nor prompted by the subconscious or the environment. Researchers believe they come down purely to biology.

Tom Uhde, at Detroit Medical Center in Michigan, believes the attacks are a reaction to some internal biological cue and that sleep panic may be quite distinct from panic disorder. If he is right and they have the most biological form of the panic disorder 'bug', further study of this group could advance our understanding of the biology of panic.

Sleep panic can lead to a vicious circle. Just as panicking behind the wheel can stop you driving, people who have had sleep panic attacks become reluctant to sleep, frightened in case they have an attack. Sleep deprivation is known to provoke panic attacks, so as they get more tired, they become ever more likely to have an attack.

Relaxation seems to be the trigger. Even through the day, those with sleep panic are most likely to have an attack while taking it easy. They tend to move around less in their sleep than others with panic disorder, possibly showing that they are over-relaxed. Most people with panic disorder move around in their sleep far more than non-anxious people and only ten years ago researchers believed this was because panic disorder keeps up arousal levels. This may be true, but Uhde now says that increased arousal at night may protect some patients from sleep panic. Increased arousal, rather than causing panic attacks, might in some situations prevent them.

Many questions remain to be answered. For most of those with panic disorder, uncomplicated by sleep attacks, increased arousal is bad. They panic once they have passed a certain threshold of arousal, and because of their increased vigilance they are always nearer that threshold than non-anxious people. But those with sleep panic might be affected in a totally different way, showing all the same outward signs, but with different causes. The cascade of events that leads to a panic attack could start somewhere else altogether, although, for the moment, this is speculation.

Both over-arousal and under-arousal set the scene for panic attacks, Uhde says. It is as if vulnerable people need to maintain their arousal levels within a narrow band and any move too far in either direction can lead to a panic attack. The discovery of sleep panic might have far-reaching consequences for research into the biology of panic attacks. People with panic disorder are an extremely mixed bag, their problems caused by a tangled mass of different factors. This makes straightforward research difficult. But those with sleep panic *could* represent a homogenous group for whom biology is the overriding cause.

They may, after all, turn out to be a discrete group with a distinct underlying cause, who perhaps require different treatments. Imipramine is widely used for panic disorder and seems to be as effective for people with sleep panic as it is for the others. But benzodiazepines increase relaxation. Theoretically at least, they could make sleep panickers worse, if by calming them down they increase the chance of a panic attack.

Alternatively, people with sleep panic might have the most severe form of panic disorder, with the most extreme reactions. In that case, scientists studying this group may get results that are relevant to everyone with panic disorder.

## *Caffeine High*

Coffee, tea and cola are such ordinary drinks that we tend to forget they contain a fully fledged drug. Caffeine is so much a part of our lives that we give it little thought beyond recognising that it can help stimulate us on occasion.

However, researchers have used it to demonstrate the importance of arousal. If anyone is given enough, even those who have never had a panic attack, they will become anxious and some will panic. The effect is stronger in anxious patients and they need less caffeine before they start feeling panicky. Some people with panic disorder have fewer attacks if they cut down their caffeine intake.

Scientists cannot agree whether caffeine has a general effect, pushing everyone closer to the panic threshold, or whether it has a specific effect and attaches to receptors in the brain called adenosine receptors. These receptors could stimulate the locus coeruleus, mentioned earlier, and trigger the release of noradrenaline, which fires the body's defences and prompts the typical pale skin and racing heart.

Drinking coffee and tea is not all negative. We like the taste, they are sociable drinks and they can help us through a boring afternoon at work. Even primary school children get a boost from a dose of caffeine. Eight- to twelve-year-old children in Minnesota were asked to drink cola with its taste disguised so they could not tell whether or not it contained caffeine. Those given the drug had a longer attention span in laboratory tests and were more dextrous in simple exercises moving pegs around a board. The children said they felt less sluggish after drinking the cola, but also more anxious.

Caffeine, like all drugs, deserves respect. It does not give the pleasure surge associated with addictive drugs that are inhaled or injected. It is legal and unlikely to send someone on a downward social spiral. But it is a potent drug, especially for those prone to panic. Students studying for exams or professionals pushed up against a deadline may drink coffee to help them work through the night. Sleep deprivation and caffeine can be powerful panic triggers where there is underlying anxiety. In combination they have a far more dramatic effect.

## States and Traits

Scientists' aim in researching the biology of panic disorder is to find the root cause of disturbances. A difference certainly exists between the locus coeruleus and noradrenergic system of those with panic disorder and those without. But it remains unclear whether the super-sensitive nature of the system is a cause of the disorder, or the result of living with it for many years. The complexity of the brain and the fluidity of so much of our biology make it fantastically difficult to tell whether such a difference led to the development of panic disorder or whether the body's reaction to panic disorder altered the workings of this system.

A key question, then, is whether people who go on to develop panic disorder had glitches in their noradrenergic system before their problems started. And whether, once it is successfully treated, the glitches disappear. This puzzle has been dubbed the State–Trait debate. Is the glitch a result of someone's current panic disorder, their state? Or is it a more permanent trait, which existed before they became ill and will persist long after they are treated?

One way of exploring the locus coeruleus/noradrenergic system is to look at the downstream effects. Neurones from the locus coeruleus project to the hypothalamus, which in turn controls the pituitary gland. When the locus coeruleus fires, the pituitary gland releases growth hormone. It can be made to fire artificially by the drug clonidine. According to Uhde, one of the most consistent findings in panic disorder is a blunted response to clonidine. People with panic disorder release far less growth hormone than others after receiving the same dose of clonidine. In this way, the growth hormone response to clonidine can be used as a research tool, and Uhde's team started using it as an indirect test for panic disorder.

As the work progressed, it became apparent that growth hormone itself might have a more direct role in the development of panic disorder. Scraps of evidence linking growth and anxiety have been around since the 1960s, when doctors first noted that children who had suffered severe physical or emotional abuse were often exceptionally small. They have so-called psychosocial short stature and if moved to a safe and caring environment, many have catch-up phases of growth. Presumably the children's anxiety at critical stages of development quite literally stunts their growth.

Other children lack growth hormone for purely physical reasons and untreated, would fail to grow. If they receive synthetic hormone while still young enough to respond, they should reach normal adult height. This is growth hormone's most obvious effect, but increasingly researchers are finding a variety of functions which are boosted by growth hormone, even in adults. It increases the ratio of muscle to fat. More controversially, it may pep up the immune system. It may even slow down aspects of ageing and increase libido. Much remains to be proven but it seems likely that in future adults with reduced growth hormone levels might benefit from a boost.

Adults with anxiety disorders, in loving environments and with no physical reason for under-production, may nevertheless have a reduced supply. Given that they have reached their final height, the impact of low levels of growth hormone cannot be measured. But drugs like yohimbine, caffeine and another – called growth-hormone releasing factor – all trigger its release. And people with panic disorder consistently show a reduced response, with lower growth hormone levels, to them all.

Is this a possible cause or a result of anxiety? It could be that people with panic disorder have lived so long with such high levels of noradrenaline that they have become insensitive to it. If receptors for noradrenaline are permanently over-stimulated they may become downgraded and unable to give a normal response to a specific trigger. Equally, the receptor for growth-hormone releasing factor in the pituitary gland – the last link in the chain before the growth hormone is released – could be faulty. The drugs ultimately exert their effects through growth-hormone releasing factor, so any abnormality at this last hurdle would prevent the full response.

Uhde moved to a breed of dogs, pointers, for a model of panic disorder. Normal pointer dogs are affectionate to people, whether they know them or not. They run to greet visitors and explore their surroundings if taken somewhere new. They will take food from anyone and will interact with other dogs. Children happily play with these dogs – and want to take them home afterwards.

Nervous pointers look and behave like normal pointers

until about nine months old. Then they become anxious on contact with humans. When someone approaches, they retreat and cower, either freezing or salivating. They tuck in their tails and adopt a submissive position. If they are taken to a new cage or field, they will not explore and may stand frozen for up to three hours if left alone. These nervous dogs dislike eye contact so much that researchers find it best to approach by walking backwards toward them.

Pointer dogs are bred by researchers for this trait. After a few generations, they breed almost 100 per cent true. Nervous pointer dogs produce nervous puppies and confident dogs produce confident puppies. The nervousness in these dogs is caused by their genes and no matter how researchers manipulate their environment, they cannot prevent a dog from developing nervousness.

After the work in humans with clonidine, Uhde's team was aware that growth hormone and IGF-1 (insulin-like growth factor 1) may be involved in anxiety. When researchers went back to the dogs, they were struck by their size. The nervous dogs were definitely smaller than the normal dogs. They had not noticed this originally, but there was a clear difference in the size of the dogs. All dogs had equal access to food and the researchers do not believe the difference was caused by the amounts they ate.

Growth hormone stimulates the release of other growth factors, among them IGF-1, and this proved easier for the researchers to monitor. When they measured the dogs' IGF-1 levels, they found a link with fearfulness. The more nervous the dog, the higher its IGF-1 level. The height and weight of the dogs was also associated with its IGF-1 level, but the best predictor was fearfulness. This is further evidence of the intricate involvement of the growth hormone/IGF-1 system in the development of fearfulness as a biological trait. It could be that genes involved in fearfulness alter the workings of the growth hormone system and reduce levels of growth hormone and IGF-1. This would in turn reduce the animal's final height and weight.

The implications of the finding are still being worked on. Synthesised growth hormone is sometimes given to short children who lack a normal supply. Treatment is stopped as soon as they reach normal height and is assumed to protect them from the possible social problems of being short. However, they remain deficient in growth hormone.

A colleague of Uhde's, Manuel Tancer, wanted to find out how these children fared psychologically as they got older. It had always been assumed that they would be relatively stable people, since their lack of growth hormone had been corrected at a critical stage. However, he found extremely high rates of panic disorder and social phobia among them and consequently suggested that the shortage of growth hormone may itself have an impact on psychological development.

These are preliminary findings, years from being translated into new treatments. There are still many questions to be answered. Uhde wants to explore the direct involvement of growth hormone in a series of studies. One is to compare children whose lack of height is associated with their growth hormone levels with others short for different reasons. But it is intriguing that some fully-grown adults with anxiety and phobias might in future be given growth hormone to treat their anxiety disorder.

This work, as ever, is unlikely to apply to everyone suffering from anxiety disorders. It contradicts Kagan's assertion that those with anxiety disorders have a tendency to be taller than average, but both agree in their prediction that such people may be more than usually slender.

It also fits with other theories about panic disorder. Panic disorder may be the final common outcome for many different malfunctions. It is as if each of the systems feeds into a great river. If any one of the tributaries suddenly triples its flow, the river banks may not hold and panic disorder may result. So whether the initial problem is hyperventilation, arousal, growth hormone, serotonin or another neurotransmitter, the final outcome could be panic disorder. Different people would have different underlying causes. Panic disorder might just be a description of what happens when any one of a list of things goes wrong and we could be clubbing together a whole group of different conditions under the same name.

This is important when we come to treat it with drugs. To re-use the earlier analogy, the drugs currently used for panic disorder might be trying to alter flow in the great river rather than in one of the smaller tributaries. The drugs may not act specifically on the problem's original cause. Some drugs probably act on some tributaries but we have no way yet of matching the drug treatment to the patient.

This may change with our advancing knowledge of the biology of panic. In future, doctors may be able to tell those with panic exactly where their problem lies, and a drug treatment specific to that problem will be prescribed. A group of people with panic disorder might all be taking different treatments. And, hopefully, all will have their panic brought under control.

## CHAPTER 5

# Behaviour

## *Triumph Over Phobia*

Andy left work early to get to his meeting on time. He took the train from Manchester Piccadilly to Crewe, where he had time to kill. He met some friends in a hotel lobby, as arranged, and they gave him a lift from there. They arrived just before 7 p.m. at the meeting hall in Manchester, fairly close to the station Andy had just left.

Andy was neither lost, nor on some mad mystery tour. He had simply been completing his homework for the week. His mission had been to travel outside a self-imposed safe zone within central and south Manchester.

The meeting was the local Triumph Over Phobia (TOP) group, where he had been a regular for the past eight months. His first panic attack was on a train in 1985 and he had dreaded journeys ever since. 'The main problem with me is apprehension. I was completely relaxed arriving at the station but in the last ten minutes before the train was due I started having flashbacks to past episodes.

'Six or twelve months ago, before I started to tackle the problem, that would have swept me away and I would have had a panic attack on the back of that flashback effect. But coming to the group, and reading up on it as well, helps you to forestall a lot of the anxiety that would otherwise have gripped and controlled you. It is just a matter of practice, trying to put yourself through it as much as possible.

'I have started to control it. The worst that happens now is that I get manageably stressed. Before I was unmanageably stressed. I would have to get up, walk up and down the train, lock myself in the toilet, go to the bar, do something. Whereas now, for the past few weeks, I have never gone beyond the level where I clench slightly.

'I was most apprehensive about the car journey back. I was worrying about us being stuck in a traffic jam and not being able to get out. But even that stress was manageable. And, as usual, it was not as bad as I expected in the end.

'It is completely irrational. But if you keep putting yourself in those situations, de-sensitising yourself to it, it does seem to work itself out.'

Andy's detour was made at the suggestion of Diana, the group leader, a lively, friendly, cheerful woman, herself a former agoraphobic. Her confidence and warmth belie a very different past: only a couple of years ago she was confined to her home, too nervous even to answer the door or telephone.

Diana is married with a young daughter and now runs a successful family business. But she was dogged by agoraphobia from her mid twenties to early thirties. It started when she was on holiday with her husband John. Getting out of bed one morning, she apparently had a fit. She was unable to breathe, her lips turned blue, she collapsed, passed out and was rushed to hospital. It looked like epilepsy.

Once home, medical investigations drew a blank on epilepsy, but not before Diana had lost her driving licence for two years. It was later established that she had had a panic attack. Over the next few weeks she had several more and her world started to shrink. Increasingly reluctant to leave home, within three months of the first attack she only saw John, and her mother, who lived close by. She could not face anyone at work, but was able to continue going in for a couple of hours in the evening after everyone else had gone. She lived only a minute from the office and knew she could bolt home should the warning signs of an attack begin.

Fear of having a panic attack took over as the main problem. She did not want to go out in case she had an attack; later she stopped opening the door in case it happened while she was talking; then the telephone became off-limits, for what would someone think if they were speaking to her and she had an attack? The walls closed in. Diana is, and had been, a sociable woman and her new life was alien to her. Some mornings she would wake up and weep for hours at the thought of another day of it. She was so low at times that she half-heartedly considered suicide.

John was bewildered by her change but remained supportive. He had thought she was going to die when she collapsed in the hotel room. Although agoraphobia took away part of Diana, he had not lost her altogether and was thankful for that at least.

Diana was also fortunate to have an exceptionally helpful GP. He arranged for a psychologist to visit her for weekly sessions which helped slightly, but a thirteen-week course was insufficient to make progress. Next came counselling. Diana walked into the room where two chairs and a box of tissues were set out and laughed out loud at the idea that she was expected to cry. The sessions explored her childhood experiences, at home, at school and with friends. Diana was happy as a child and her relationship with her parents remained good; she absolutely could not see what her upbringing had to do with her agoraphobia. Counselling got nowhere. Her GP offered medication, which she tried once so she could attend a family funeral some distance away. She got through the day, but felt disembodied, as if she were watching herself from over her shoulder, so she never took the tablets again. Almost desperate to help, her GP, who lived round the corner, came round to check her blood pressure in the morning before she got up. It was normal. It seemed that agoraphobia was simply outside the GP's remit.

Diana turned to alternative therapies and spent a fortune on aromatherapy, which she loved but it did nothing to help her agoraphobia. Homeopathy came next and then anything else she saw advertised. Nothing worked.

The turning point finally arrived when the health visitor came round to carry out a standard check on her young daughter, Leona. Like the women described later in chapter 8, Diana felt much better while pregnant and breastfeeding her baby. Unfortunately, her agoraphobia re-surfaced soon afterwards and by the time Leona was eighteen months old Diana's inability to leave home was clearly restricting her daughter's childhood. The health visitor put Diana in touch with the charity, Triumph Over Phobia (TOP), and hence on the road to recovery. It took Diana several phone calls with the group leader to convince her to go along, but she was determined to get better and eventually went.

TOP runs self-help groups throughout the country under the guidance of Professor Isaac Marks, the doyen of behaviour therapy. It follows Marks's book, *Living with Fear*, to the letter. Group members, like Andy, are encouraged to expose themselves to the things they fear. Over weeks and months they carry out increasingly difficult tasks, and if all goes well, their anxiety fades.

When they arrive for the first time, people are asked to list the phobias they want to address. They do not list every fear – only the phobias that impinge on their lives.

Most groups are between six and eight people. They meet weekly, usually for two hours, to catch up on each others' progress and to set 'homework'. Meetings typically start with a discussion of the previous week's homework. Everyone describes what they have done, and how anxious they felt before and during the exercise. They learn relaxation techniques to help them deal with anxiety and panicky feelings.

Homework is tailored to the individual. Joanne, who is afraid of spiders, had read a book on them and touched all the pictures that week. She had also walked across the Humber bridge, alone, on a windy and rainy day, to address her fear of heights. Jamie has social phobia and had managed to strike up several conversations with women in the office that week. He had also taken out membership of a gym and gone twice.

The group divides into twos and threes to negotiate tasks for the following week. They each take away four tasks, and when two sound rather similar, Diana lumps them together, suggesting taking on something else. The homework is meant to be demanding – Marks suggests that people spend two hours a day on it. Few group members are this devoted to their own recovery and Diana's approach is to alternately praise her group for past achievements and push them further.

Everyone completes a fear questionnaire once a month in which they grade their fear of situations, from zero, which means they never avoid them, to twenty, which means they always do. They can trace their progress in these questionnaires, especially helpful after a bad week. A bad week now might mean a fear level of four when walking across a bridge; whereas three months ago, it could have been seventeen.

Behaviour therapy the TOP way is carried out according to a few strict principles. One is that you must stay in the feared situation until your anxiety starts to subside. If you venture bravely into a supermarket or on to a bus, feel your anxiety rise and promptly leave, it could make things worse. Another is that the homework has to be done often, preferably every day. The longer you leave it before having another go, the harder it gets. If you take tranquillisers, which is discouraged, take them a few hours before you set off.

Lasting friendships are made at group meetings, sometimes unlikely ones. Diana's group includes people who often have nothing in common except their phobia. These can sometimes be oddly complementary. A woman with a fear of chemical contamination attended the group along with another who had a fear of dirt. One would leap in the shower if she splashed even a drop of bleach on herself. The other cleaned her house continually till her hands were raw. In an attempt to persuade her new friend that bleach was harmless, the second woman memorably drank some – not to be recommended. However, both women eventually left the group much improved.

Sometimes a bit of imagination is needed to devise tasks to get at the phobias. Diana periodically arranges a group outing to Manchester Airport, where there is something for everyone, she says: every conceivable type of lift, glass and metal, fast and slow, big and small; escalators; crowds; cafés and shops. The claustrophobics can try out the lifts; social phobics can sit in cafés; one woman normally unable to talk to strangers was sent to all the car hire booths to find the cheapest deal.

Having a group to report back to can be a big incentive, not unlike going to Weight Watchers. Many people with phobias could devise their own programme, but sticking to it without support is difficult. Diana says that when she had agoraphobia, she always had a better week when she went to the group, even if she did not feel like going and said little when she got there.

Despite the camaraderie, Triumph Over Phobia is a course, not a club. People should not become so comfortable they are reluctant to leave. Joanne is a case in point. She has been attending for almost a year and has made huge improvements. When she first arrived she brought a shopping list of phobias, which she has worked through steadily. She will not be turfed out of the group come week fifty-two, but she is being encouraged to think of moving on, and to use her last few weeks to tick a few more things off her list. Joanne has been helped immensely, but is not cured. She has flown to Paris for a weekend with her boyfriend and started to play in a local orchestra. But she has not returned to work and – though she feels much better – some of her problems remain. She does not want to leave the group.

In the two years since Diana started her group, there have been several others like Joanne, but only three people who gained nothing. The treatment on offer was just too brutal for them to contemplate.

Diana, by contrast, embodies the success possible for this technique. She was highly motivated to improve, partly so that Leona would not miss out on a normal childhood. She religiously set aside two hours a day to do her homework. She laughs now at the memory of herself walking the streets of Manchester in the dark. Like so many people with agoraphobia, she felt less conspicuous then. (Chapter 9, on light and electromagnetism, discusses possible reasons for this.)

Sometimes she finds it difficult to understand when people in her group do not turn up or cannot find the time to do their homework. When ill, nothing was more important to her than getting better. As soon as she started making progress, and her anxiety began to fade, her success drove her to achieve still more. It took eight months before she was ready to leave the group, and she vowed then that she would never again turn down an opportunity through fear. This vow has opened up a whole new world. Her work with Triumph Over Phobia has introduced her to celebrities, it demands public speaking and involves meeting journalists. Past interviews are pasted up on a wall in a room at home. Socially, her horizons are wider than ever; she has flown and is soon off on a cruise with a girlfriend – without John. She is back to work at full strength and Leona is now a chirpy, outgoing five-year-old.

She has been warned that she could face setbacks, but she has felt herself for so long now that she cannot imagine it. She is sure that she is totally cured.

## Self Help

Behaviour therapy has been around for decades and TOP's approach is only one of several. Early versions often started with de-sensitisation in fantasy. People listed a series of feared situations and graded them from the most to the least difficult. They had to imagine one of the easier situations while remaining relaxed and once they could manage this, they moved on to something more difficult. Eventually, they went out and tried it in practice. So-called flooding in imagination was similar except that it started with the most difficult tasks. People continuously imagined the worst scenes they could think of, for an hour or two, while the therapist chipped in with suggestions of other frightening situations or details to make the imagined situation worse.

These methods declined because of convincing evidence that real-life exposure is both quicker and more effective, though imagination techniques are still sometimes used for those too anxious to confront the situation they fear.

Behaviour therapy is now being repackaged for the twenty-first century in the twin garbs of efficiency and technology. The primary requirement is to cut down on therapists' time, and high-tech approaches make this possible.

Early exposure therapy was led by a professional, and phobics were accompanied as they took steps towards facing their fear. Marks then demonstrated that results were as good without the therapist. The therapist's role became one of supervising the choice and grading tasks for people to attempt alone. Many ask a spouse or friend to accompany them, especially the first few times. But the final goal is to complete the task alone and it is not a good idea to become dependent on anyone.

The move towards self-directed exposure has led to an explosion in the ways in which exposure therapy can be delivered. Computers, books, telephones, even just a postal system, can provide sufficient contact with the therapist. It requires an ever-shrinking amount of professionals' time and schemes once based on weeks of treatment have been condensed into hours, especially for specific fears, say of spiders or snakes. Many psychiatrists find they can direct programmes as effectively by telephone as by meeting up face to face. Video telephones might increase the effectiveness of this approach in future. Even without such technology, psychiatrists like Susan Shaw, a colleague of Marks's, receive homework diaries by post and see a certain percentage of patients only rarely.

Irrational fears are so widespread in the community that minimal intervention has much to recommend it. No health system could cope with demands from such a large number, and few would have qualms about trying off-the-peg treatments in the first instance.

Marks has developed a computer system based on his book. People key in feared situations, read an explanation and select exposure tasks. The following week, they record achievements and select more tasks, just as the book would recommend. Marks claims that use of the book or computer system can be every bit as successful as face-to-face sessions with therapists. In a study, agoraphobics were randomly assigned to receive *Living with Fear*, or obtain exposure tasks directly from the therapist or from the computer. The three groups improved at a similar rate and to a similar degree. From being habitual avoiders who felt extremely anxious in certain situations, all improved in the first three months. At six months, they did not avoid situations, had no phobic panic and only slight residual anxiety.

This progress required little time from therapists: none for those reading the book, just over an hour for those using the computer and over three hours with those they saw directly. But conditions in such a study are not typical. People using the computer or book could see a psychiatrist at any stage of treatment. None choose to do so, but the offer alone could make a difference. Assessments throughout the study may also have boosted determination. Other research into self-exposure programmes has confirmed that they help, but not always by as much as Marks's group found. On rare occasions they have been of no help at all, but this appears to be when agoraphobia is extremely severe. Drop-out rates are a consistent feature of such demanding therapy and even in Marks's study more than one in eight did not complete the course.

Shaw has since adapted the computer program to allow users to select their own goals rather than having to choose from a list. This may improve further the computer-directed exposure programme and she says their advantage over self-help books is that they feel more personalised. She is currently assessing its acceptability. A group of people with various phobias use the computer for twelve sessions lasting anything from thirty minutes to a couple of hours. In the end, a combination of computer and therapist may prove to be most effective – a twelve-session therapist-directed course for agoraphobia could, for example, be reduced to four sessions with the therapist, plus unlimited use of the computer.

Shaw's system sends people out into the world to do their homework but other systems allow homework to be carried out on site. Researchers at Leeds University are devising a program dealing with the fear of spiders. The computer shows a series of pictures of spiders; people attempt to sit out their fear and then press a button to call up a soothing landscape on screen.

Virtual reality may help in the future. It can create life-like pictures that allow users to move around a scene – a café by a water's edge, a plank leading out to water, or a suspension bridge across a gorge. The scenes change as the user walks and the overall impression, if not totally natural, is convincing. A system like this for fear of heights has been devised by an American company, Division Limited. The company says that nine out of ten of those who try the system are then able to meet goals in real life, such as driving across a bridge or going up in a glass lift while looking at the view.

Interestingly, these most high-tech of approaches hark back to earlier ideas of desensitisation before exposure to the real thing. Researchers will continue to come up with ever more imaginative ways of delivering behaviour therapy but its core remains the same.

## Do It Yourself

Kay was struck by a casual comment on a radio programme. 'Everyone smiles when they look at themselves in the mirror,' said the presenter. Kay did not. She could not risk catching a glimpse of her teeth. She had learnt to floss without looking in the mirror, and put on lipstick without showing her teeth. If she was talking beside a mirror, she would turn away from it.

Kay had been afraid of all things relating to the dentist for as long as she could remember. She avoided walking past dentists' premises and would even get upset if she heard strangers talking about it on the tube. 'I would get the whole anxiety thing. I would feel my heart going fast, I would be flushing and feeling strange. I would have to walk away from them. I suppose if anyone had asked me about it, I would have known it was irrational, but most of the time I didn't think about it at all. I hadn't spoken to anyone about it.'

She had 'super-duper avoidance' for twenty years, but then, perhaps inevitably, got serious toothache. 'At first I just ignored it. It wasn't that I was frightened of going to the dentist, more that it was impossible. It didn't cross my mind as a solution.'

Kay, a solicitor, was then thirty-eight years old and decided the time had come to address her phobia. 'It wasn't just that I started to have problems, it was also that I felt more mature. I have two children and I thought that if I could go through labour twice, and cope, I could do something about it. I also knew that I did not want to transmit this to my children.'

But she had no idea where to start. She assumed her GP would dismiss it, and had not spoken about her fears to anyone. Finally, she told her husband. 'He was a bit shocked, he hadn't noticed that I never went to the dentist. But I knew that once I had told him, he wouldn't let me get away with it.'

Kay then embarked on her own version of exposure therapy. Her experience, though ultimately successful, demonstrates the value of a properly structured programme and support.

She found a dentist who specialised in dealing with dental fears and made several visits before even looking at the equipment. She agreed to be sedated to have some work done. 'It was extremely unfortunate. I was told afterwards that I had been tearing my hair out and behaving as if I were insane. I woke up in tears trying to fight them off. It was partly the anaesthetic's fault, but things weren't explained to me clearly. Maybe that was difficult because I was behaving like a zombie. Agreeing to the sedation was a big thing for me and I had thought if I got over that I would start to feel better. But I felt awful afterwards. I was confused, upset and generally negative about everything.'

She went back three more times but, when a filling fell out, decided to look for another dentist. This meant taking a day off work and visiting three more dentists before finding one she liked. Several months on, she has a dentist she trusts, her own teeth are in order and she makes regular appointments for her children. She successfully addressed her own phobia but feels the experience was more traumatic than it needed to have been.

'I wish I had talked through my feelings before I started. I felt so distressed and confused when I was doing it. It was a complete nightmare, my whole emotional balance was disrupted for months. I wasn't sleeping. I couldn't concentrate, I was very ratty and couldn't think about anything else much of the time.

'I was motivated to sort it out and it became a challenge. It was almost experiential, I wondered what would happen. Would I have a nervous breakdown or would I survive? In some ways I got quite a high – I was doing the unthinkable and was still alive. I felt I was some great heroic person, dealing with a huge monster. At other times I felt life wasn't worth living.

'My family suffered. My husband tried to be supportive but had no idea where I was coming from and he found it difficult. I would walk out of the house and he wouldn't know whether I was coming back or not. I would charge around the house, or walk round in tears and I was horrible to the children. I had no patience and was depressed. The other day our Nanny said something about my daughter's bad behaviour at the end of last year. I was shocked. That was when all this was going on, and I knew it was because of me. I felt I had done something dreadful.'

## Fight Fear With Fear

So what happened? Why did Kay's dental appointments and Diana's late-night walks put paid to their phobias? What exactly was going on as they subjected themselves to their ordeals?

Unusually, behaviour therapy is more readily accepted by the lay public than by scientists. The logic of getting straight back on the horse you have just fallen off makes sense to us all, as does 'practice makes perfect'. We encourage children to try out new experiences even if they do not want to; we all have to psych ourselves up sometimes, to go to a party full of strangers or to make a public speech. Fight fire with fire, goes the old adage, and fear with fear in the case of phobias.

It is not a sufficient explanation for the experts. Complex, ingrained phobias like agoraphobia and social phobia are debilitating anxiety disorders often associated with a string of other problems. Social phobia is often coupled with alcoholism; agoraphobia, with problems in the family and possibly, as discussed in previous chapters, genetics and chemical disruptions in the brain. It may be triggered by the death of someone close. Asking someone whose problem was prompted by intense grief to walk into a feared situation has no compelling rationale.

Behaviour therapy grew out of a body of research conducted largely on laboratory animals. Learning theorists gave animals rewards, usually food, and punished them with unpleasant shocks or noises. I. P. Pavlov, mentioned in chapter 1, was an early learning theorist in the 1940s whose dogs famously heard a bell before receiving food. They soon learnt that the bell meant food and started to salivate on hearing it, even if no food arrived.

Ten years later, scientists had concluded that anxiety or neurosis was a learned behaviour. They used reward and punishment systems to understand how and when animals avoid situations. They demonstrated that animals can be taught to avoid situations, but also that they can be subsequently trained *not* to avoid these same situations.

Psychologists of the day, such as Hans Eysenck, were keen to start applying such results in the clinic. From the scientists' point of view, agoraphobia was a golden opportunity. There is no shortage of people with the condition and their behaviour can be measured in a straightforward way. There is no need for subjective interpretation of complex psychiatric questionnaires – someone who is housebound at the start of treatment and can walk to the shops by the end is a success.

The serious nature of agoraphobia was also appealing. The scientists reasoned that if they could demonstrate success with agoraphobia, the most intractable of anxiety disorders, then they would have proved their point and could go on to work with other such disorders.

But simply training people to change their behaviour was a radical departure from most treatments of the time and it was controversial. Mainstream therapy following Freud was psychodynamic and analytical, with therapists attempting to explore and understand underlying reasons for fear. Some of their explanations are now easily ridiculed. A fear of spiders was linked with paranoid tendencies and problems with sexual identity. Phobics were thought to be in denial of this, transferring their fears on to the spider which represented an orally devouring and anally castrating mother.

Such a bagful of psychiatric problems, according to

psychoanalysts, required months or years of intense analysis. Simply exposing highly vulnerable people to the supposed object of their fear was extremely dangerous. At worst, the exercise would intensify the phobia, precipitating calamity and exacerbating the underlying problems. At best, any benefits would be superficial and the fear would return in another guise. Someone treated for a fear of spiders might become afraid of snakes instead, substituting one set of symptoms for another while the underlying problem remained unaddressed.

In fact, symptom substitution did not happen. Early results in the behaviour clinic were encouraging, attracting attention and support. Adults who had been afraid of spiders all their lives were cured. Not just helped, but cured. Phobias simply disappeared and many people had no further problems. Success was achieved within a time frame barely conceived of beforehand.

More recent work suggests that successful treatment of one phobia actually reduces other fears rather than increases them. Jack Rachman, a British psychologist now working at the University of British Columbia, is a leading cognitive therapist. He studied a group of people afraid of both spiders and snakes and found that if one fear was treated, for many people, the other diminished as well.

The treatment in Rachman's study was given by cognitive therapists, but the idea of a helpful knock-on effect may be relevant to other therapies. If treatment of one phobia changes thinking or behaviour in a way relevant to another phobia, it is likely to undermine that second fear as well.

## A Bitter Pill

While symptom substitution as an objection to behaviour therapy has been roundly quashed by experience, other concerns have proved more enduring. One is the harshness of behaviour therapy. Behaviour therapy alone, as practised by Triumph Over Phobia, is not for everyone. Walking into the situation you fear above all others and deliberately exposing yourself to something you have avoided for years demands tremendous courage. It is simply too much for some.

The task, as seen by the phobic, is immense. Stroking a dog might feel a bit like spending an afternoon in a lion's cage. Even if the phobia severely restricts your life, even if you believe that a single-session course could put everything right, it is not necessarily acceptable. The phobia and the fear are far bigger than outsiders can imagine.

Many barriers have to be overcome before phobias can be addressed. As discussed in the introduction, community studies have found that up to 40 per cent of Western populations live with fears but only a tiny proportion ever seek professional help. Even today, phobias come with feelings of shame and inadequacy which can make free discussion impossible. It takes great courage to come forward for treatment, and the prospect of exposure therapy might be horrific enough to put many people off altogether.

Self help can take time and dedication. Diana is a terrific advertisement for the potential of the technique, but it took her eight months at two hours a day to get better. That is almost 500 hours of extreme stress, or about three months' full-time work. Not surprisingly, many of those who embark on a course of behaviour therapy do not complete it. Cognitivists (discussed in the next chapter), who use behaviour therapy in a slightly different context, claim to be able to get much quicker results, sometimes in a single session and usually within six. They say it is not realistic to expect everyone with a phobia to be as determined as Diana.

Levels of motivation may be as relevant to success as levels of fear, and the two are not always closely related. Change is always difficult, and never more so than when an entire family has become organised to accommodate agoraphobia. An agoraphobic may be severely impaired but find the prospect of change unappealing, perhaps unconsciously even weighing up the benefits of the condition. It may prevent them doing a job they do not like, or force family members to give them more attention. Super-caring spouses can make things worse. If a husband takes care of all the domestic chores because he cannot bear to see his wife suffer, he may unwittingly make her worse by reducing the need for her to do anything. Alternatively, a husband may be unconsciously reassured if his wife is at home all day. Perhaps it means that the house is well ordered and well kept; perhaps he feels it prevents her having an affair.

In this way, agoraphobia can allow family members to manipulate each other. These scenarios are often implicit, with those involved unaware of what is going on, but it can incapacitate, while providing little incentive for improvement.

At the other end of the spectrum, many are extremely afraid but highly motivated. Shaw says that some of her patients are quite obviously terrified. They take a taxi to the door of the building, bolt inside and spend the entire consultation holding their spouse's hand and shaking. Many subsequently do extremely well on a programme of exposure therapy.

## Behaviour Therapy – the
## Easy Way?

One way to make behaviour therapy more palatable may be to combine it with medication. Advocates of this approach claim that drugs can reduce anxiety sufficiently for exposure therapy to become possible. The drugs scale down the fear and the exposure stamps it out. However, opponents such as Marks are suspicious of the use of drug therapy to dampen anxiety.

Phobias often get worst during a spell of depression and few professionals argue with the need to treat depression with drugs. Depression needs to be controlled before people can benefit from most phobia treatments. The argument focuses on drug treatments for anxiety in the absence of depression. In *Living with Fear*, Marks suggests that if benzo-diazepines are taken, it should be in small doses several hours before going out to complete an exposure task. If they are taken just before leaving home to ride on a bus, the journey must last until the effects have worn off. Otherwise, the fear may simply return. More recently, he has suggested that using drugs may be counterproductive if too much faith is invested in them.

A group of phobics received behaviour therapy and drugs combined. Eight weeks after their treatment started, they had improved substantially. Researchers asked what they attributed their gains to. Those who believed their improvement was due to the drugs relapsed when they were assessed again later. Those who put the improvement down to their own efforts continued to improve after the end of the active treatment.

Marks insists that one of the overriding advantages of undergoing behaviour therapy alone is that the results are more enduring. A study carried out in both London and Toronto compared people given exposure plus drugs with those given exposure plus a sugar pill placebo. After eight weeks, those taking the active drug had improved slightly more than those on placebo. Many studies are wound up at this stage and the take-home message is that drugs help. But this study continued. The drugs were gradually reduced and ten months later patients were assessed again. This time, those who had not received the active pill were significantly better off than those who had.

The study is complicated because the drug, alprazolam, was given at high doses where lower doses might have been less damaging to progress. However, Marks's point is that exposure therapy is best undertaken neat.

But Michael Liebowitz, from Columbia University in New York, believes that medication may reduce the drop-out rate and help more people through behaviour therapy. Current research into social phobia appears to back him up. The drug phenelzine shows impressive early results. After six weeks, those taking the drug were significantly better than those who received just cognitive and behaviour therapy. Overall severity, social anxiety, impairment, avoidance and anticipatory anxiety were all reduced. But by week twelve, those on cognitive and behaviour therapy had started to catch up, and a few months after the treatment's end, those who had received cognitive and behavioural therapy were more likely to have remained well. Such studies suggest that cognitive and behaviour therapy combined are best for those who can bear it. Others might consider adding a drug therapy.

Translating the results from clinical trials into ordinary

clinical practice, however, is far from straightforward. Clinical trials may discern the best approach for a particular phobia, but doctors are treating people, not conditions. If someone is not happy to try the approach suggested, there is no reason why they should. Many people with phobias have definite views on the approaches they will consider. Some feel strongly that they do not wish to take medication. Others are equally vehement that drugs are the only approach they will consider and they are not prepared to embark on any kind of behaviour therapy. These views are often formed after previously unsuccessful attempts to address phobias. The choice of treatments is increasing and nobody should have to accept an approach they have little faith in.

## Courage Under Fire

Another concern is that exposure therapy may change behaviour but leave some residual fear. People like Joanne, who can now walk over bridges and drive long distances alone, have been helped but not cured by behaviour therapy. Psychologists have claimed that some are left feeling afraid and anxious, even though they have changed their behaviour. A man with social phobia may be able to go to parties and give presentations after a course of treatment, but still harbour the doubts and fears that originally led to his phobia, they say.

Fear and avoidance make natural partners. As a rule, we do not enjoy being frightened and, given the opportunity, avoid such situations. This observation was at the heart of our understanding of phobias for many years. Mowrer's two-stage theory, first published in 1939, stated that an unpleasant or painful event caused fear, which is a reaction to pain and which also motivates behaviour. The fear in turn prompts people to avoid the unpleasant event in future and, in so doing, reduces their fear.

The theory held sway for over a decade. Problems began to arise in the 1950s when laboratory work suggested that avoidance was more persistent than predicted. If animals receive a mild electric shock when they take food from a particular ledge, they learn to avoid that food, as the theory suggested. However, they continued to avoid it long after the shock had been stopped. Mowrer's theory predicted that the avoidance would tail off.

The work thus raised doubts about the theory's universal application. A multitude of common examples of uncoupled fear and avoidance were even more compelling.

Avoidance without fear is common in everyday life. A student who misses lectures or fails to write up essays may be lazy, rather than afraid. He or she may avoid work in order to reduce their effort, rather than their fear. Someone who refuses to go to lengthy films or operas may be avoiding boredom, not fear. An office worker who never eats in the staff canteen might simply be avoiding bad food. It is not unheard of, even for someone with agoraphobia, to be extremely avoidant but surprisingly unafraid. The avoidance may have started at a stressful time and increased even as the fear lessened. Sometimes the fear can disappear altogether, leaving behind incapacitating behaviour, but no particular anxiety.

Fear without avoidance is typical of some phobias. A young man with social phobia who regularly goes to functions, avoiding only the biggest or busiest, may be extremely afraid but not avoidant. It can be more obvious in the dramatic circumstances of war; many of our heroes and heroines were extremely afraid. The Second World War provided psychologists and psychiatrists with an extraordinary opportunity to study fear and avoidance. In his excellent book, *Fear and Courage*, Rachman describes the uncoupling of fear and avoidance among airmen and civilians. A study of 6,500 airmen found that between a third and a half were afraid on almost every mission – not surprisingly, since at one period in 1942 RAF bomber crews had only a 10 per cent chance of surviving a full tour of operation. It rose to a scarcely encouraging 50 per cent in other campaigns. Still the airmen were willing to return for a second tour, but their willingness depended on various factors. Pilots were more willing than other members of the crew; unmarried men were more willing than married ones. Injuries were not a deterrent nor was fear. Almost three-quarters of those pilots who claimed to have been afraid on every mission were willing to return.

In the exceptional circumstances of war, at least, fear can exist independently of avoidance. Of course, airmen were highly skilled, competent and confident young men, but ordinary people on the ground were equally robust under fire. Experts expected that air raids on cities would lead to widespread panic and hysteria. Cities were evacuated in preparation, and special psychological clinics were set up. Doctors in Liverpool trained volunteers as auxiliary mental health workers but none of this was necessary.

People were resilient. They grew used to air raids and developed remarkably few phobic reactions or psychiatric disorders. British, Japanese and German civilians all displayed similar bravery. They went about their business as best they could. They were repeatedly subjected to frightening situations and got used to it. In fact, city dwellers subjected to regular and frequent raids displayed less fear than those in rural areas where attacks were much more sporadic.

Wartime conditions are exceptional but these studies may be useful in our understanding of fear today. Fighting for king and country might seem highly motivating but people were not usually driven by national pride or the sweep of history. Family, friends and peer pressure, then as now, regulated motivation. Careful questioning of soldiers revealed that ideological convictions were irrelevant for the vast majority. Social bonds and approval within small groups were far more important in helping them through unimaginable stress. They would risk injury and death rather than exclusion from society. The airmen were proud to belong to small, tightly bound units, within which flying and fighting were the only acceptable ways to behave. Fear and avoidance often do go hand in hand but they *can* disengage themselves, and this is important to anyone with a phobia today.

Behaviour therapists claim that uncoupled fear and avoidance is rare in normal circumstances. The argument developed out of laboratory work when scientists noted that if they repeatedly blocked animals' escape from an unpleasant shock, the animals eventually stopped trying to avoid it. The fear remained, though, and if anything, increased. Many clinicians and psychoanalysts thought the same would happen to people. If forced to face their most dreaded object or situation, their behaviour towards it might change, but their feelings, perhaps, would not.

A key difference between the laboratory work and exposure therapy in humans, say behaviour therapists, is that animals spent a comparatively short length of time in fear. The success of exposure therapy in the clinic depends on sitting out the worst of the anxiety. That can mean half an hour, an hour or even more. Fear levels rise initially, but then level off and fall. This is one of the principles of behaviour therapy and therapists stress that it is quite possible that if people walk away from a situation before their fear starts to subside, like the animals, their fear may increase as a result.

Behaviour therapists therefore argue that as long as the rules are followed, avoidance will diminish and so will fear. But arguments about residual fear have rumbled on over the decades and have been taken up more recently by cognitive therapists. They say that fear and avoidance, though usually linked, can sometimes disentangle themselves and leave people who have rid themselves of their phobia with an unpleasant constant anxiety.

Many factors appear to influence how closely related fear and avoidance remain. Experimenters suggest that the more intense the fear, the more likely it is that fear and avoidance move in tandem. The more frightened you are of a situation, the more likely it is that you will want to avoid it. You are also more likely to have a physical reaction to it, perhaps a wildly beating heart.

Effective treatment should address all aspects of fear. Exposure therapy ostensibly concentrates on behaviour but proponents believe that its effect generalises over time. Shaw believes that fear and avoidance move together. If someone changes their behaviour and starts facing up to whatever they are frightened of, then their fear will be reduced. They will still remember that they once had a severe phobia and will carry memories of their unease but, she says, they are left with just the memory rather than the feeling itself.

Concerns about uncoupled fear should not be overstated. Even if behaviour therapy only works to reduce avoidance and does not abolish fear, it can still give many people their lives back. In practice, most have their fear diminished, and for some it disappears altogether. But if behaviour therapy is set to become more available, the point needs to be studied further. If the therapy gives people back the ability to go to work, to socialise or to carry out domestic chores, it is surely a good thing. But it would be far better if they were able to do these things free from excessive anxiety.

## Why Does It Work?

None of the work described has given an adequate description of how and why behaviour therapy works: this is because we do not know. The therapy, which has consistently shown its value in the laboratory, in the clinic, in self-help groups, among readers of books and users of computer systems, is based on successful practice, not theory.

It is effective but it does not *always* work. Marks admits that 3 per cent of people can follow the rules exactly without their fear starting to subside. Others less convinced about behaviour therapy put the failure rate higher. The 'non-habituators', as they are called, make an intriguing exception and not even Marks will hazard a guess as to what is happening.

He is, however, impatient about the debate over how behaviour therapy has its effect. The only point to note, he says, is that it usually works, and should be more widely used. In his opinion it has suffered because it is so cheap and because nobody stands to make any money out of it; if it could have been patented by a pharmaceutical company, it would have taken off years ago. The Department of Health recently agreed to fund a full-time officer for TOP, but this is a tiny investment compared to the money routinely pumped into drug development.

To an extent, Marks's anger is unfounded because most modern treatments for phobias now include behaviour therapy. Cognitive therapists, whose work is described in the next chapter, originate from a different tradition within psychology and a couple of decades ago would have given behaviour therapy short shrift. Now, they – like most researchers – accept that it works even if we do not know why. Cognitive therapy is so often combined with exposure, that it is usual to talk about CBT, or cognitive and behaviour therapy. Behaviour therapy is routinely combined with drugs, counselling and hypnotherapy. Few successful treatment packages are complete without it.

The efficacy of behaviour therapy is no longer disputed, whether within a package or on its own. Laboratory work has provided few insights into why it works and nor, so far, has neuroscience. It remains frustrating for theorists that one of the most successful tools in the treatment of phobias is based on so little evidence. But anyone with a phobia can be reassured that we know, at least, that it works.

## CHAPTER 6

# Cognition

## *Testing*

Stephen's first attempt at behaviour therapy landed him in casualty. He set off with his therapist into the city centre, had a lot to drink and visited many different loos. But he was completely unable to use any of them. His phobia is of public toilets and, as the afternoon wore on, he became increasingly uncomfortable and desperate. The harder he tried, the less he was able to urinate. Finally, he panicked and went to the local hospital, where fortunately, he was able to use a quiet toilet.

The episode set him back for months. 'For the next few days after this I was really depressed and the next few months were not easy. It really sent me down,' he says.

Stephen, who is now twenty-two, has lived with his phobia since he was twelve. He was bullied at school and in one crucial incident a group of older boys came into the toilets where he was and started banging away at the cubicle doors. When he moved to the upper school, his phobia got worse. 'I remember the first day going to see what the toilets looked like. The cubicles had doors which did not go right to the floor, and after one go at using them, I ended up not using them for the rest of the time I was there. I took to waiting until I got home.'

The phobia was not addressed, and by the time he got to university Stephen was anxious and having regular panic attacks. In his second year, he moved in with a family, but could not use the toilet with anyone around. Eventually, he dropped out of his course and went home. He started doing voluntary work, helping out with a local school, and taking a computer course. All were within easy reach of home, so that he could go back at lunchtime to use the toilet. With this restriction, a full-time job was out of the question.

He recovered from the failed attempt at behaviour therapy, and embarked on a course of cognitive therapy. This focused on his beliefs and self-esteem. His fear is a type of social phobia, at its heart the fear of being embarrassed. Many, many sessions later, he has come to believe that it is highly unlikely – rather than extremely likely – that anyone in a public toilet would show any interest in him, let alone attempt to ridicule him, and that even if they did, he would cope. 'Before, I would have totally lost it. I would have collapsed and maybe ended up in hospital. Now, I think I could always just walk out.'

With a new set of beliefs in place, he is attempting behaviour therapy again, and with increasing success. 'It was necessary to sort out my ways of thinking before going on to do the behavioural work,' he says. He still has some way to go but has started using public toilets. His confidence is much improved, he has made new friends and has been accepted on to a foundation course to re-start at college.

Cognitive therapy is not always such a long haul. Stephen's phobia required a change in thinking at a fairly deep level. No matter how determined he was to urinate, if he was having negative thoughts his body would not let him succeed.

But cognitive therapists say that even deeply ingrained and

complex phobias can be treated in a single session. Professor Paul Salkovskis from Oxford University was so confident of success that he agreed to be filmed treating a woman with long-standing agoraphobia. They had never met before the documentary (*The Mind Field*) was made.

Salkovskis broke down her fears into a series of simple – but false – beliefs and systematically challenged them. For example, the woman, Tracy, felt as if her legs were going to give way in the much-dreaded supermarket. They were weak and shaky, as if they could not possibly support her full weight. Salkovskis asked her to stand on one leg, which seemed ridiculous. If two legs could not keep her upright, how could one? But she found she could stand on one leg without falling over and came to think that perhaps, after all, her legs were sturdy enough.

She was feeling breathless, as if she could not take in sufficient air. Her chest felt tight, she was breathing quickly but none of the air seemed to reach her lungs. Surely she was about to suffocate. Salkovskis suggested she hold her breath for twenty seconds. Dubiously she agreed and, against all odds, as she saw it, she did not suffocate and die.

Her racing heart surely meant she was about to have a heart attack, but the therapist asked her just to concentrate on some relaxation exercises for a couple of minutes. Afterwards, her heart seemed to have slowed down. Yet you cannot relax away an impending heart attack.

Salkovskis systematically challenged Tracy's assumptions about what would happen in the supermarket. At the end of the afternoon she was amazed and delighted. She had stayed in the shop for almost two hours and felt extremely pleased with herself. She had neither collapsed nor fainted and previous certainties now seemed less likely. She had been sure that should she collapse, nobody would have helped her and she would have been left lying on the ground. She had been afraid of dying of a heart attack and leaving her children with nobody to look after them. In a single afternoon, she took a shaky but big stride towards recovery. Her progress continued over the next few months until she was able to shop in central London, not without qualms, but without running away.

Tracy is in her early thirties, and has two young children. She has had agoraphobia for several years and the strain of it broke up her marriage. She had sought treatment before but nothing ever worked. Her GP was sympathetic, but seemed to have nothing to offer her but tablets which she did not want to take. The treatment offered by a behavioural therapist, which meant going to the supermarket and staying there for thirty minutes at a time, scared her almost witless. She was unable to continue and been discharged because of a 'lack of motivation'.

In fact, to an outside observer, cognitive therapy looked very like behaviour therapy. She first discussed her fears with Salkovskis and worked out what she wanted to achieve. She then went with him to the local supermarket, somewhere she had not been for years. She stuck out her fear, was thrilled with her success and continued to improve over the ensuing months. But the core of cognitive therapy is teasing out and testing false assumptions. When Tracy stood on one leg to test her belief that she was about to collapse, or held her breath to see whether she was on the verge of suffocation, these elements of her treatment were pure cognitive therapy.

Behaviour and cognitive therapies are usually combined in clinical practice but they grew out of quite separate traditions and the thinking behind them is crucially different.

Cognitive therapy is an independent school of thought, not a variation on the theme of exposure.

When behaviour therapy was first introduced in the 1950s, its success rate astonished psychologists and psychiatrists. Previous therapies had devoted months or years to exploring the underlying causes of fears. Exposure turned thinking about phobias on its head. It became apparent that a single hour of exposure to the object was far more likely to help many patients than hours of talking. As discussed in the last chapter, Professor Isaac Marks at the Institute of Psychiatry pioneered and promoted much of the early thinking. But, as mentioned previously, he has not speculated on *how* it works, believing theory to be unhelpful. Exposure works and that is enough, he says.

Most other researchers now reject this stance. They want to know why and how it works in order to help them improve on it and to hone other phobia therapies. Salkovskis and David Clark at Oxford University have long been in the vanguard of thought about behaviour and cognitive therapy. Both trained as behaviour therapists, later developing a more cognitive way of working. Since the late 1970s they have been building an understanding of how exposure works. Its success is indisputable but it is a blunt instrument and aspects may be unnecessary, unhelpful or off-putting. If the essentials of the therapy can be distilled out, treatment may become increasingly effective and efficient. It may be shorter and less painful and perhaps become acceptable to more people.

Cognitive therapists believe that exposure works because it disconfirms false or unhelpful beliefs. Some beliefs are quite easily challenged by exposure. At the one-day course for people afraid of flying, described in the introduction, approximately one-third of the delegates had never seen inside a plane. Their fears were based on false assumptions about aircraft interiors or the flight itself. One man thought he would have to bend double to walk down the aisle because there would be no headroom. Another believed that the noise of the engines would be deafening for the whole of the flight. The course climaxed with a short trial flight and could hardly fail to help these people overcome their fears. Their beliefs were disproved by a single trip in the air, and when these beliefs were at the root of their fear, the phobia disappeared as well.

Other false beliefs are harder to challenge. A further third of the participants on the flying course had different and more complex reasons for their fear. Some thought that the anxiety they would undoubtedly experience could itself harm them. Some had fears more akin to claustrophobia, agoraphobia or social phobia. People with agoraphobia might focus on the vast expanses of space around an airborne plane. Those with claustrophobia might dwell on the moment when the aircraft door is finally closed, after which they have no way of getting out. People with social phobia may be afraid that they will make a fool of themselves on the plane by getting nervous and losing control.

The cognitive model of emotion devised by Professor Aaron Beck in the mid 1970s stresses that our interpretation of events is the key to understanding our emotions. A passer-by may or may not believe that the event warrants our reaction but that is irrelevant. We are made anxious, according to this theory, by our beliefs about what is happening, not necessarily by the event itself.

If, for example, a masked man broke into my office, put his hand across my mouth and a gun to the side of my head, I would be frightened. If the man was on the run from the police, looking for a hostage and somewhere to hide, I would be right to be frightened. If, however, a friend was playing a sick joke, the gun was plastic and the incident being filmed for the company's Christmas video tape, my fear would be misplaced. But I would remain extremely frightened until I knew I was safe and my initial reaction would be identical regardless which scenario turned out to be correct. The anxiety would be created by my belief that the masked man was prepared to hurt or kill me.

In Beck's model of anxiety, he assumes first that the more likely we think a threat is, the more anxious it makes us. Similarly, the more terrible the consequences of this threat, or the higher its cost, the more anxious we get. But our anxiety is reduced if we think we will be able to cope with the threat and if we have lots of rescue factors, such as friends and family, who we think will help us out.

Beck came up with an equation which appears highly mathematical but is intended just to aid our thinking on anxiety. It is:

$$\text{anxiety } \alpha = \frac{\text{perceived probability of threat} \times \text{perceived cost}}{\text{perceived ability to cope} + \text{perceived rescue factors}}$$

The equation helps explain some apparently anomalous observations. Some people with phobias accept that the thing they most fear is unlikely to happen but remain terrified of it. According to Beck, that is because it would be appalling if the incident were to happen: its cost is high. Tracy may have believed there was a one-in-ten chance of her having a heart attack in the supermarket and dying. That is bad enough, and anyone would want to avoid situations with a 10 per cent chance of death. But her fear was multiplied further by concern for her family. Dying was not as bad as the thought that there would be nobody to care for the children.

Tracy was hindered still further by her belief that if she felt ill she would simply collapse and be unable to help herself. She was sure that passers-by would think she was mad and step over her as she lay on the ground. She would be left for dead even as she choked on her own vomit. She assumed that she would neither cope nor be helped and this, according to Beck, pumped up her anxiety.

Similarly, Stephen's improvement took place in parallel with changes in his thinking. Once he started believing he was highly unlikely to be ridiculed, and that he could cope by walking away should this ever happen, using a public toilet became far easier.

Beck's view makes sense even when people do not have an anxiety disorder. Salkovskis himself never had qualms about flying until his daughter was born. Since then, his thoughts turn to her whenever a plane hits turbulence. It is not going to crash, he is sure. But *should* it crash the consequences are too awful to contemplate. The event is remote but its cost is immense. He would never see his daughter again, would not see her grow up, would not be there for her when she needed him. He has not become afraid of flying, but is no longer the blasé passenger he was.

## Spiralling Out of Control

Beck was trying to simplify the complex web of fears and beliefs that underpin and maintain anxiety. David Clark wanted to build on this and see how a general understanding of anxiety can be applied in practice to phobias. Phobias can linger for years, sometimes lifetimes, and Clark set out to discover why fears persist. He came to the conclusion that phobias are the result of not one, but three vicious circles, all feeding into and being fed by fear. These circles relate to, respectively, selective attention, arousal and safety strategies.

Selective attention means we notice the thing we fear, and having noticed it, become more afraid. Consider a typical university department, littered with unattended briefcases. It sounds totally benign and it normally is. However, if staff have just received a memo about a terrorist threat to the department, these briefcases suddenly become very obvious and threatening. Selective attention transforms unexceptional briefcases into potential bomb carriers, and appears to multiply their numbers tenfold.

In the same way, anxiety creates threats in environments most would consider harmless. Most people rarely see a spider, yet to someone afraid of spiders, the world is populated with them. They see every spider within an improbable radius. They notice spiders' webs and all the dark, dusty corners where spiders could lurk. Awareness and constant vigilance stoke anxiety, setting up a vicious circle. The more the anxiety, the more vigilant people become, the more potential threats they notice.

To the social phobic, the threat comes from other people. They fear rejection, humiliation, scorn. At parties, they scan the face of the person they are speaking to for the tiniest hints of a negative reaction. Someone suppressing a tiny yawn probably had a late night but the social phobic takes this to mean their companion is bored. Anyone who glances across the room is desperate to get away. If someone should drift off to fill an empty glass, they are probably regaling half the party with tales of the dreadful nerd they just got stuck with.

These interpretations are, of course, counterproductive. The more attention the anxious person pays to the minute reactions of their companion, the less they can think about what they are saying. Which probably makes them less coherent, charming and witty than they would otherwise have been. Those with social phobia very often say nothing at all for fear of failure. Alternatively, they may dominate the conversation as an inappropriate reaction to cover their fear.

For those with specific phobias, the threat is at least external. It is possible to avoid lifts and distance yourself from spiders. Both have costs. Not using lifts can cause problems at work and avoiding spiders can mean sacrifices if picnics or holidays in the countryside are out of the question. But it can be done. The agoraphobic, however, is stuck. Here the threat is internal and therefore ever-present. Someone with agoraphobia is afraid of collapsing, having a heart attack, maybe suffocating. They selectively attend to bodily functions, scanning themselves for fluctuations and slight abnormalities. The problem with this is that bodily functions fluctuate constantly. It is quite normal to be slightly out of breath at the top of a long flight of stairs. If we leap up from a chair or bed, it is not unusual to feel a bit dizzy. And if we are nervous our heart rate will often speed up.

Clark's second vicious circle is arousal. Someone with agoraphobia enters the supermarket already feeling nervous. Fear is a potent physiological stimulus and a surge of adrenaline can have many effects, including speeding up the heart rate. Careful scanning detects this instantly, proof that, as expected, they are about to have a heart attack. This increases the fear, increases the heart rate further and confirms the heart attack theory.

Similarly, the person feeling breathless assumes they are about to suffocate and tries to compensate by breathing more. They start to breathe unnaturally rapidly, forcing air in and out of the upper bronchioles, so that it hardly reaches the lungs where it can be properly absorbed. They feel increasingly out of breath, further proof that they are about to suffocate.

Agoraphobia is closely tied in with panic and cognitivists believe that the crucial step between self-scrutiny and full-blown panic is catastrophic misinterpretation of these symptoms. Being slightly out of breath is equated with suffocation; dizziness with fainting; and a pounding heart with an impending heart attack.

It is difficult to over-estimate the horror of this experience. Someone who believes they are having a heart attack is getting repeated confirmation from their body that they are right. They are convinced they are about to die. Their emotions are identical to someone really having a heart attack. We sometimes talk glibly about panic, but the immediacy and intensity of it can be very like the experience of someone taken suddenly, possibly fatally, ill.

Panic and the vicious circle of arousal is most obviously associated with agoraphobia but it also applies to social phobia. Anxiety at parties can make sufferers blush, sweat, shake and draw attention to themselves by looking obviously uncomfortable – the very situation they wished to avoid at all costs. Even a vague, quizzical glance can set anxiety soaring and exaggerate all the physical symptoms.

Arousal and catastrophic misinterpretation of symptoms can form a powerful and important loop in pumping up anxiety. But there is yet another mechanism. Clark elicited the third vicious circle by re-examining the obvious. Many people with phobias face the object of their fear repeatedly, but never become less afraid. If anything, their fear increases. People with a fear of heights and loathing of bridges may still have to cross bridges and drive over flyovers every day. Many with agoraphobia still have to go out to work, and those with social phobias cannot always avoid giving presentations or going to office parties. Behaviour therapy is based on the prediction that repeated exposure will reduce fear. So how come these fears appear untouched by exposure?

Clark questioned people with phobias in great detail and discovered that many have developed personal strategies that they depend on to prevent calamity. These so-called saving strategies come in a variety of guises, some more obvious than others. The first is out-and-out avoidance. Someone afraid of the supermarket shops at the local corner shop instead. They never go near the supermarket so their belief that they will collapse at the door remains untested. Others rely on escape. They regularly set out to go shopping, pick up a basket and start walking down the aisles. But the moment they start to feel dizzy or panicky, they put down the basket and walk straight out of the shop.

Still others stick it out, despite feeling grossly uncomfortable. Clark discovered that many have more subtle ways of dealing with anxiety, such as holding tightly on to the trolley at the first hint of unease. This seems harmless enough. But Clark discovered that people who believe in and use these so-called saving strategies, like those who leave at the first sign of trouble or will not go near the supermarket at all, leave their false beliefs untested.

If they believe they will faint, and that clinging on to the trolley will prevent this, they leave the supermarket thinking that they had a near miss. If the trolley had not been there, they think, then they really would have fainted. So their underlying belief that 'supermarkets cause fainting' remains intact and may even be strengthened by the experience.

Cognitive therapists believe these saving strategies may underlie many of the failures of behaviour therapy. Behaviour therapy works, say cognitivists, by getting people into the situation they most fear and demonstrating to them that nothing terrible happens. If they are using subtle strategies to, as they see it, minimise the danger they are in, their original belief and fears remain unchallenged. Apparent success in not succumbing to the danger is attributed to the saving strategy.

Safety behaviours are completely normal if viewed from the agoraphobic's point of view. We all know roads are dangerous and when we are driving we stop at red lights. This is a safety behaviour. If we were to drive on we believe we would be in danger of having an accident. Most people who stop at red lights have never tested out this belief. But they stop without thinking when the lights are on red. Likewise, someone with agoraphobia may believe they are in danger of having a heart attack if they do not leave the supermarket instantly. They have no wish to question this belief because if they are right, they have a heart attack. It therefore takes great courage to walk into the feared situation without the perceived armour of a safety behaviour or two.

## Squaring the Circle

Cognitive therapy aims to halt these vicious circles by challenging the assumptions at the root of someone's fear. This can take considerable imagination. Putting a stop to selective attention might require the therapist to demonstrate quite conclusively that the situation their client is looking out for is not dangerous and simply will not harm them. A colleague of Salkovskis was treating a woman who would not enter a room if the doors and windows were closed. Careful questioning revealed that she was afraid all the air in the room would get used up. The therapist left the room, closed the door and sprayed some air freshener in the corridor. The client could soon smell it, and this simple demonstration was enough to convince her that doors are not air-tight and she would not suffocate in a room with closed windows. Afterwards, her phobia improved.

Slowing the arousal circle requires cognitive therapists to address the misinterpretation of symptoms. Many people, like Tracy, believe that a racing heart means an imminent faint. Another colleague of Salkovskis asked a woman with this belief to think of other times when she had been aware that her heart was beating wildly. The patient remembered when she heard she had won on the pools. Her heart had gone berserk but it was thrilling. It was not unpleasant, and she had not even thought about fainting. Therefore a racing heart was not necessarily a bad thing and need not signal an on-coming faint. In one session her beliefs were challenged.

More simply, Tracy thought she was going to faint, but also knew that her heart was pounding. Her therapist got her to agree that people faint when their blood pressure drops too far. Her pounding heart would mean that her blood pressure was raised. This in turn would be enough to stop her fainting. She eventually reasoned that, however awful she felt in the supermarket, she would not faint because her blood pressure would be raised, not lowered, by her fear.

People like Tracy are often taught relaxation techniques

to get them through the experience of deliberately walking into a dreaded situation. Simple exercises can reduce anxiety but they have many other more subtle benefits. They give people something to concentrate on when they start to get uneasy. Many feel they are more in control of the situation if they have something to do. But over and above all this, relaxation can help patients challenge their own beliefs. Someone who is starting to panic, starting to feel they are suffocating or having a heart attack can force themselves to carry out relaxation techniques. If this makes them feel better, they can be sure that they are not seriously ill. The knowledge that they are panicking, and are not about to die, is an immense relief and the spiral of catastrophic thoughts that lead to a full-blown panic attack is halted.

People must also be encouraged to forgo or directly challenge their saving strategies to stop the third vicious circle. Exposure to their feared situation will only have the desired effect if they give up subtle measures of self-protection. If someone always walks close to a wall when they feel off balance, they are encouraged to walk away from it. Someone who tenses their legs in an attempt to prevent them giving way is persuaded to relax their legs deliberately, even to stand on one leg. This should prove to the patient that their legs are, after all, quite sturdy.

Someone who has relied for many years on a safety behaviour may be virtually unaware of what they are doing. They may have internalised their own, peculiar, precautions. Just as we give no thought to stopping at the red light, so the agoraphobic may think nothing of their dependence on the supermarket trolley. Cognitive therapists take great pains to tease out these strategies so that people can be persuaded to abandon them.

For anyone who has practised subtle saving strategies for years, cognitive therapy may be more demanding than behaviour therapy alone. These patients might be quite willing to go to a shopping centre on request; it is the loss of the trolley or nearby wall that creates the problem. But without adding this element to the exposure therapy, the patient's real beliefs will have been left untested.

Cognitive therapists, then, address patients' selective attention to the thing they fear, their arousal and catastrophic misinterpretation of bodily symptoms, and any saving strategies they might employ to help them get through a situation. This is usually combined with some exposure to the phobic object or situation and the growth of combined cognitive-behaviour therapy in practice is a testament to its success.

## Disgust

The fear of imminent death through heart attack or suffocation carries a certain logic, but what about the fear of slugs, cockroaches or worms? None could conceivably offer a bodily threat to a human, yet all are commonly dreaded. How does the cognitive therapy apply here?

Some therapists have come to believe that a model based entirely on fear is insufficient. They say that many phobias are instead based on disgust. Disgust phobias lead to exactly the same behaviour as any other – the fear and avoidance can be extreme – but the root cause may be a sensitivity to disgust which needs to be addressed for the problem to be properly resolved.

Disgust phobias are typically directed towards small animals and creatures which rarely, or never, mount an attack on people. Three separate studies by different research groups all found snakes, cockroaches, rats and spiders in the top six of a list of the most intensely feared animals. All score highly on disgust ratings.

Many features of spiders are reported to be the cause of the fear. Their movements, ugliness, legs, colour and so on. Graham Davey from Sussex University, whose work was discussed in chapter 2, says this probably means the real reason for the fear is hidden. He believes it may be disgust.

Davey compiled a list of common British animals from suggestions made by undergraduates. He then asked students at an Open University summer course to rate their fear of these animals. Fear of indigenous animals in the UK is not easily explained by fear of attack, and Davey suggested that disgust is more plausible. Women scored consistently higher fear ratings than men, possibly because men are less willing to admit to fears. But Davey found that gender differences were only apparent in fear of animals thought to be disgusting. Wasps and bees are often feared but rarely found disgusting and there was no difference in men's and women's fear of these insects. However, women were significantly more sensitive to disgust than men.

Disgust is not fear and we all know whether we feel afraid or repelled. But Davey believes it can underlie and lead to fear. Working with student Katie Webb, he set up a series of videos aimed at manipulating volunteers' moods. One group saw a violent video, one saw scenes from post-mortems and hospital operating theatres. The third group watched views of an aerial landscape. Before and after seeing the video, volunteers were asked to rate their fear of twenty-eight different animals. The animals were classed as predatory, non-predatory, repulsive or neutral.

The neutral video appeared to relax viewers and they gave lower scores to all groups of animals after watching it. But the effects of the other videos were specific. Watching the violent video increased fear of predatory animals such as lions and sharks but reduced fear of all other animals. Watching the disgusting video, however, increased fear of both the non-predatory animals like rats and spiders, and of the repulsive creatures such as slugs and maggots. This suggests that sensitising someone to violence heightens their fear of predatory animals. Sensitising them to disgust increases fear of snails and other slimy creatures.

Davey believes that disgust sensitivity and violence sensitivity can both lead to fear. The fears they produce are different, but both may be strong enough to be classed as phobias. Which means, he says, that phobias related to high disgust sensitivity levels should not be treated as just phobias. The underlying sensitivity to disgust must also be addressed if the phobia is to be eliminated once and for all.

Disgust is an obvious reaction. Imagine finding a maggot on your plate after you have started eating; watching the cat vomit up a half-digested mouse on the kitchen floor; having to use a dark and particularly smelly toilet. Your facial expression is unmistakable. You wrinkle up your nose, push your tongue forwards and out, your mouth may gape open and you exclaim in disgust.

The distinctive disgust expression is almost exactly the same as the look on a newborn baby's face when a sour or bitter taste is put on its tongue. This similarity may be because disgust grows out of the food-rejection response of babies. The expression is designed to prevent you from inhaling or swallowing any of the undesired substance. Wrinkling up your nose blocks off smells. Moving your tongue forward stops you swallowing and your gaping mouth lets you get rid of anything already in there. Perfect for straightforward food rejection.

Charles Darwin, in *The Expression of the Emotions in Man and Animals*, said that the movements round the mouth in extreme disgust are identical with those preparatory to vomiting. He went further:

> The suspicion arises that our progenitors must formerly have had the power (like that possessed by ruminants and some other animals) of voluntarily rejecting food which disagreed with them and now, though this power has been lost, as far as the will is concerned, it is called into involuntary action, through the force of a formerly well-established habit, whenever the mind revolts at the idea of having partaken of any kind of food, or at anything disgusting.

He added:

> We can see that as a man is able to communicate by language to his children and others, the knowledge of the kinds of foods to be avoided, he would have had little occasion to use the faculty of voluntary rejection, so that this power would tend to be lost through disuse.

He suggested the expression we make when disgusted is a physiological leftover from the time when our predecessors were able to vomit at will but which remains useful in warning youngsters away from dangerous foods. The expression is near univeral and Darwin described it among 'wild Indian tribes of North America', 'a young Hindoo', 'high-caste natives' and Fuegians. He also recognised cultural influences on disgust.

> In Tierra del Fuego a native touched with his finger some cold preserved meat which I was eating at our bivouac, and plainly showed utter disgust at its softness; whilst I felt utter disgust at my food being touched by a naked savage, though his hands did not appear to be dirty.

Many of these ideas have been been re-visited recently, and may help our understanding of some phobias.

Darwin focused on adult disgust towards foods, but its scope is wider. It includes animals, parts of animals and animal products such as mucus or faeces, but also other people's dirty habits or sexual behaviour, sometimes our own body shape or eating habits. Whatever the stimulus for the disgust, though, the facial expression remains the same.

The transition from food rejection to disgust starts at around two or three years old and the prompt is usually faeces. Before that, babies and toddlers will put slugs, beetles – virtually anything into their mouths. Quite how young children develop the reaction is not known but parents are an important influence. The strength of a child's disgust reaction to a disgusting object or situation is similar to that of their parents', and this suggests the reaction is transmitted down through the family. Children may copy their parents' expressions in empathy and then come to associate the reaction with the stimulus. Both verbal and non-verbal communication is likely to be important. In fact, parents usually instigate potty training at the age children are starting to develop a disgust reaction. They may be taking advantage of their children's growing awareness and possibly moulding this developing reaction at the same time. Older children learn to associate features of faeces and mucus with a range of animals. If a snail looks slimy, or mucus-like, that may be enough to stop them wanting to touch it.

Paul Rozin and A. E. Fallon pioneered thought on disgust. They say it consists of nausea, avoidance, a reluctance to let a disgusting object into your mouth and a fear of being contaminated by it. This makes sense when considering the food-rejection response but is harder to connect with a fear of certain animals. Davey believes it comes down to associations we make with the animals. The disgust reaction to rats or cockroaches, for example, may be because we link them with the spread of disease. In the case of rats this is understandable, since they have been linked with disease since the Black Death. Rats still carry numerous pathogens and a recent study of rats trapped on farms in England and Wales found that they carried twenty-three different bugs, most of which could cause disease in humans.

Animals such as snakes, lizards and worms, however, are not thought of as great reservoirs of disease. They may simply remind us of the primary disgust foci of faeces or mucus. They look slimy, therefore they disgust. Other animals, maggots for instance, signal infected or unfresh food and indirectly are associated with disease.

One possible function of disgust is that it has evolved to help animals keep themselves and their nests in sufficiently hygienic conditions. It would discourage them from eating or drinking contaminated food and drink and may have reduced the incidence of infection and illness. Rozin suggested that disgust may also be a means of cultural communication, a quick way of signalling that objects or animals are to be avoided for cultural reasons.

Both explanations are highly speculative but disgust does appear to be universal. According to Davey, only Polish and Chewong do not have a word for the idea of disgust. Samoans do not distinguish between hate and disgust but some cultures differentiate categories rather precisely. The Ifalukians of Micronesia distinguish between disgust linked to decaying food and the disgust of moral outrage.

Stimuli for disgust have broad similarities across the globe and tend to include body waste products and decaying animal matter. More specific stimuli differ across cultures, though, and where in Korea it is acceptable to eat dog meat, in the UK this would be regarded as totally disgusting.

This all suggests that disgust, like fear, is usually adaptive, though not always. Disgust sensitivity plays a role in several important anxiety disorders. When it becomes linked to something inappropriate, like body shape or eating, it can precipitate eating disorders. An exaggerated sensitivity to disgust can give an overwhelming and constant fear of contamination. In obsessive-compulsive disorders, this in turn drives sufferers to constant cleaning rituals, often repeatedly washing hands or cleaning the house.

Those most sensitive to disgust are most likely to fear animals we consider disgusting such as spiders, rats and snails. By contrast, fear of lions, tigers and predatory animals is entirely unrelated to disgust levels. So if many phobias of small animals are mediated by disgust rather than fear, this suggests that cognitive therapy should address underlying disgust sensitivity. A Dutch group found that those phobic of spiders, regardless of how they acquired their fear, were more resistant to standard behaviour therapy if they had high disgust sensitivity levels. Davey suggests that those who are highly sensitive to disgust would benefit more from desensitisation to disgust and contamination in general than from direct treatment of the phobia. Failure to deal with underlying disgust will leave the phobia free to return and even to spread to other animals.

The idea of phobias spreading has a familiar and rather unconvincing ring. The same argument was used in the early days of behaviour therapy. It was assumed that if someone was exposed to spiders, for example, they would simply transfer their fear to snakes. But in practice, this symptom substitution did not happen.

The disgust scenario is more complicated because people extremely sensitive to disgust are unlikely to be simply afraid of spiders and will probably have a more generalised anxiety about dirt and contamination. Therefore they are more likely to come forward for help with an obsessive-compulsive disorder than with a specific fear. It could be that a slightly raised disgust sensitivity is relevant to dislike or mild fear of common animals. It may be less relevant to clinical animal phobias.

Davey maintains that a phobia may grow out of disgust sensitivity to become full-blown fear. Disgust-based phobias are likely to become serious and require treatment, he says. But Salkovskis believes that disgust phobias are not true phobias because disgust is not fear, is not as intense and does not have such serious consequences. Disgust is a real issue for many anxious people but he says it is rare to come across anyone whose disgust sensitivity has led only to intense fear of a small animal.

These are theoretical rather than practical arguments because in practice patients' fears are individually assessed and addressed. Disgust is most often dealt with in obsessive-compulsive disorders and researchers will have to produce more evidence before its role in phobias is properly understood.

It is, though, intriguing to consider the leap from a sensitivity to dirt to a fear of spiders. Disgust is a relatively new area of debate but it underlines the complex nature of the thought processes behind phobias. People who are afraid of snakes, spiders or rats may not readily link their fear to a higher-than-average standard of cleanliness. Where this link exists, it will lie well below the surface of consciousness.

The relevance of disgust is still controversial and consensus is some years away. Whether or not it is finally accepted, there could still be many other new areas to be explored. Psychologists are trying to take a step back from the excellent cognitive work taking place in the clinic and working to deepen our understanding of the cognitive processes underlying phobias, exploring the sort of thinking people may do without ever being aware of it.

## Risk Assessment

It is not just a question of whether the glass is half full or half empty. The depressive believes it probably has a false bottom anyway, and only contains a mouthful, whatever it looks like. Her anxious friend is too concerned about additives, E numbers and possible contaminants to even think of drinking it, whereas the optimist does not care how much is left because he is sure he can ask for more.

Any situation is open to many interpretations and all may be equally valid. Any of the drinkers may be correct but what distinguishes the optimist is that only he is happy to finish the drink. He may be wrong, he may not be able to get any more, but at least he enjoys what is there.

Cognitive therapists are interested in the different responses associated with different anxiety disorders. They do not aim to make phobics more rational, they are already as rational as anyone else in most circumstances. But therapy aims to free people from persistent ways of thinking that are causing them distress. Their original way of thinking may be quite rational, but it may also be unhelpful.

Depression, anxiety and phobias have all been linked with distorted thought processes and studies suggest that the distortions are specific to the disorder. Depressives seem to be skilled at remembering negative information, especially when it refers to themselves. They selectively remember past failures and tend to dwell on their own shortcomings. Anxious people do not do this, though there is now some slight evidence that their memories may be shaped to recall threatening information.

The findings from many studies are coherent enough for evolutionists to suggest that the distortions have grown out of natural responses to environmental threats. The anxious person's focus on threats has developed out of a need to identify and respond to potentially harmful stimuli. By contrast, the way depressed people colour their memories and pick out negative events might have arisen from the need we have always had to reflect on events that have led to failure and loss. In this way, anxiety is a forward-looking distortion and over-estimates the possibilities of trouble ahead. Depression is backward-looking and reflective, and may over-estimate trouble in the past.

Depression is also linked with a horrible accuracy in people's assessment of their influence over outcomes. They do not labour under false illusions of control. This accuracy does not usually do them any good. Acute awareness of your own inadequacies, even if objectively correct, can be less helpful than a mistaken self-belief in propelling you through social and professional mazes. In this way, non-depressed people are more likely to have biased thought processes and an unfounded belief in their control over situations.

Anxiety increases the tendency to interpret ambiguous information as threatening. If a series of words with two meanings, like dye and die, are read out loud, anxious people are far more likely to assume that the threatening meaning was intended. In ambiguous sentences, such as 'The men watched as the chest was opened', they are more likely to think of an operating theatre or post mortem, than of a box or trunk of treasure.

Phobic fears lead to specific biases, according to a colleague of Susan Mineka, whose pioneering work on preparedness was discussed in chapter 2. A. J. Tomarken showed a group a series of slides of snakes, flowers and mushrooms and after each slide either nothing happened or volunteers received an electric shock or heard a tone. These three possible outcomes were paired off equally with snake slides and the fear-irrelevant slides. Afterwards, snake phobics dramatically over-estimated the number of times a picture of a snake was followed by a shock but were fairly accurate in their assessments of the other pairings. Those who were not afraid of snakes showed only a slight bias in the same direction. It suggests the phobics assume that something bad will come out of seeing the snake. Receiving the shock confirms their negative expectations.

Anxiety may bias the focus of attention. The oft-repeated Stroop test is based on cards marked with threatening and non-threatening words, written in different colours. Volunteers name the colour. Non-anxious controls are not influenced by the meaning of the word and typically take a similar length of time to name the colour of any word. People with anxiety disorders are slower to name the colours of threat words and particularly slow when the word relates to one of their pet fears. Those phobic of injury or illness would be slowed by words like injury or coronary which might not affect other anxious people. Social threat words like criticised and failure affect all anxious people. It suggests that anxiety caused by the threat diverts resources away from naming the colour, and slows down performance.

It is as if anxious people have the mental equivalent of a permanent look-out. Their look-out will find a threat among neutral or ambiguous words. He will not be distracted by a busy environment but will instead bring in resources from other departments in the brain at the slightest hint of trouble. All of us have some sort of look-out, to warn us of impending danger. But the anxious person's look-out is tireless, vigilant, always scanning the horizon for signs of trouble.

Vigilance stokes up anxiety and prevents it tailing off when nothing untoward is happening. Once somebody has developed an anxiety disorder, distortions in their thinking keep it going.

These studies are fascinating, but they do not tell us how or why the vigilance developed. Undue attention to threatening information is linked with anxiety but either could have set up the cycle: hypervigilance increases anxiety, anxiety increases vigilance.

Beck, whose work on the maintainence of anxiety was discussed earlier, believes that problems start with distortions in thinking. He says that people with anxiety disorders pick up assumptions and beliefs early in life. These beliefs consistently colour their reactions later on. They believe deep down that the world is a dangerous place and are therefore on the look-out for threats.

Evidence that distorted thinking is ingrained backs this up. Performance in the Stroop test does not depend on how someone is feeling at the time. When drugs are given to reduce anxiety, those with anxiety disorders still pick up on more threats than others and take longer to name the colour. Long-term anxiety seems to mean long-term vigilance.

This might imply that anxious people over-estimate the likelihood of bad events, but an interesting study challenges this assumption. Randolph Nesse, whose work was discussed in chapter 2, and Richard Klaas from North Carolina, devised a list of negative events that covered everything from breaking a bone or being struck by a meteorite to catching pneumonia or being killed in a plane crash. Volunteers were asked to rate how likely it was that the average American would suffer one of these events, and how likely they were themselves to be affected.

Answers from fifty patients at an anxiety disorders clinic were compared with fifty visitors to an outdoor art fair. Nesse and Klaas analysed the results and found that both groups had identical scores. Anxiety patients did not over-estimate the likelihood of the dangerous events occuring, nor did they guess themselves to be at any higher risk than did the other group. Both groups over-estimated the likelihood of rare events, such as catching rabies from an animal bite, and under-estimated the likelihood of more frequent events, such as needing to be hospitalised.

This study demonstrates that anxious people can assess the frequency of dangerous events as accurately as anyone else and appears to challenge the body of evidence suggesting that anxious people think the world is especially dangerous.

But Nesse and Klaas say their finding may still be compatible with this idea. Their study was carried out in a risk-free environment and tested abstract knowledge about risks. Reactions may be very different when a direct threat is perceived.

Nesse and Klass suggest that cognitive defects in those with anxiety disorders only become apparent in the face of real stimuli. Then, assessments of their likely involvement and how they will react go haywire. Processing theoretical information about risk, however, poses no more problem for them than for anyone else.

This makes sense. A man with spider phobia, for example, can often joke about his fears, and acknowledge that they are unfounded. But introduce a spider into the room and that changes fast. He does not assume the spider will crawl away again. Rather, he becomes suddenly convinced that the spider is about to crawl up his nose, or lay eggs under his skin – certainly affect him in some way.

## Know Yourself

Working out what is going on in the conscious mind is difficult enough. The unconscious mind is even more inaccessible and researchers have developed a variety of experiments to examine unconscious thought. For example: two words appear on a computer screen, one above the other, and volunteers are asked to read out loud the word on top. When one of the words is replaced by a dot on the screen, they press a button. They respond most quickly when the dot (called the probe) replaces the word they were looking at so this is a way of testing where their attention was focused.

Psychologists Colin Macleod and Andrew Mathews, working in London, found that clinically anxious patients were quickest when the probe replaced a threat word. Non-anxious people were quickest when it replaced a neutral word. They concluded that the anxious direct their attention towards potential threats whereas the non-anxious direct their attention away from them. It emphasises the point that anxious people are not less rational than others, but may process information differently, so that anxiety is maintained or increased.

The same researchers set up a complicated experiment in which they measured reaction times in response to cues on a screen. At the same time, volunteers listened to a series of brief stories played into one ear. Threatening and non-threatening words were played into the other ear, though volunteers were not consciously aware of this. Anxious people had slower reaction times after the threat words, meaning that they diverted attention to the threat word.

Information about threats, then, seems to be processed differently by anxious and non-anxious people. Anxious people divert more resources away from the reaction-time task in response to the threat. This happened without them even being aware of the threat. Macleod and Mathews speculated that stimuli are processed unconsciously and then prioritised. The task has to compete with the threat.

This is an extremely challenging result. Current cognitive therapy does not require people to be tremendously articulate, but it does assume that they know what their fears are. Tracy knew and could tell people that she thought she was going to collapse, faint or have a heart attack. These fears were then systematically dealt with by her therapist. This experiment suggests that people can process information about threats without ever being aware of them and it presents a new challenge for cognitive theory and practice. The anxious person may be unconsciously scanning the environment for threats, and be influenced by things outside their conscious awareness. If so, they cannot possibly be articulating all of their fears to their therapist.

Behaviour therapy, though, was highly effective long before anyone suggested that it worked by disconfirming false beliefs. In the same way, it could be that these unconscious anxieties are being addressed by current treatments. Our understanding of the cognitive processes underlying phobias and anxiety is still in its infancy but it is an exciting area of work. If this finding is backed by further studies it will be fascinating to see whether cognitive therapists can adapt their practice to take it into account. Tackling the unconscious may be the next step.

For the moment, that is in the realms of theory and the laboratory. In practice, cognitive therapy, almost always combined with exposure, is achieving good results. The two disciplines have different origins but work hand-in-hand. To Salkovskis, this is because behaviour therapy was simply swallowed up by a more progressive field of research. Exposure therapy is at the core but all the advances are coming from the cognitive school.

As Salkovskis is first to admit, he is bound to say this. Others see it as a paradox. As our thinking about phobias becomes increasingly cognitive, treatments become ever more behavioural. Study of cognitive processes is incredibly exciting, we have more and better ideas about how and why phobias develop and the underlying reasons for the fear. The behaviour school of thought has less to offer theorists suggesting simply that people have to face what they fear.

Nevertheless, treatments that include exposure to the feared object get consistently good results. Most professionals treating phobias, regardless of any intellectual prejudice they may harbour, include behaviour therapy in their treatment package. It is a paradox which has not been properly resolved.

# Personality and Temperament

## Three-Dimensional Personalities

For six years, Elizabeth slept in the back of the car. Not from a lack of choice, nor because of rows with her husband, but because her teenage daughter had started drinking with her friends. 'I was frightened to death that she would come in and be sick,' she says.

Elizabeth is phobic of vomiting. She has been afraid for as long as she can remember. Her first memory is of being three years old and sharing a bed with her brother when her parents had friends to stay. Unfortunately, her brother was sick. Her mother came and sorted him out quickly – and then spent hours with one distressed little girl.

More than fifty years later, she is still terrified. She is frightened of picking up stomach bugs and avoids crowded places. She does not like to eat out in case meat is undercooked or hygiene poor. Her pregnancy was dreadful. She felt nauseous and had stopped taking anti-anxiety pills. 'There was no way of escaping. I must admit, I kept saying to the doctor "Let me have an abortion". I got through it somehow but it's the worst I have ever been. It's basically why I would not have any more children.'

As luck would have it, her daughter Anne was nervous

and frequently sick. Elizabeth's first husband left when Anne was only five months old and for the next three years she was on her own. It was hard. 'If the baby was sick I would leave the room. I have phoned my mother at five o'clock in the morning to get her up here to look after Anne. She was sick once while the decorator was here. He didn't do any painting, he just looked after Anne. If I was on my own and she was being sick in the bathroom, there were times when I would just come out and close the door on her.'

Anne was sick after friends' parties, after the first day at school, the first time at the Brownies. 'It was difficult, especially when she was too young to tell me she felt sick. I used to keep my distance from her sometimes if I knew there was a bug going round at school.'

Elizabeth puts her phobia down to her personality. Her father and her daughter are similar, she says. 'We all have exactly the same personality. In so many different ways we are like one person.'

This personality is driven and perfectionist. Elizabeth's father was a workaholic and, though he did not have a phobia, he had 'funny ways'. He always had his hair cut at 8.30 a.m. on a Friday morning before work. He was meticulous about cleaning the household's shoes and always changed the newspaper he used on the same day of the month. The newspaper under the dog's bowl was changed on the same day every week.

Elizabeth has always been meticulous about housework, vacuuming and dusting every room every day. She is fussy about clothes and will not dry them in the house in case they pick up cooking smells. She worked in a bank when she was younger and, like her father, had to make everything perfect. Anne, now twenty-eight, sets herself similarly high standards both at home and at work. Sadly, she was recently diagnosed as having body dysmorphobia, a constant worry about her appearance. She has a terror of finding spots on her neck and checks it ruthlessly. 'If she has got a spot, even if it is a pinprick, she sees a boil,' Elizabeth says. The phobia has caused depression and disruption, though Anne is now being successfully treated with drugs and counselling.

Elizabeth, like much of the public at large, accepts that some of us are born more anxious than others. Some are natural worriers; nervy, jumpy, extremely shy, perhaps. But doctors and scientists have been comparatively slow to take on board the idea that it may be possible to pick out these people when they are very young and intervene, possibly to change the whole course of their future.

Most of us neither want nor need doctors interfering in our lives to this extent. But emerging data reveals that certain temperaments combined with certain environments may breed serious phobias. Doctors hope that in future this research may allow them to prevent many phobias before they take hold at all.

Our knowledge of different human temperaments is both very old and very new. As mentioned in chapter 1, Hippocrates divided the world into melancholic, choleric, sanguine and phlegmatic types. These were linked with one of the four elements – earth, fire, air and water, respectively – and caused by a build up of the appropriate humour: black or yellow bile, blood or phlegm. Ideas have moved on and temperament is now seen as a combination of biology, psychology and social environment, determining how we react to the world.

Many different personality scales have been suggested. One of the best known is the Eysenck Personality Inventory of the 1970s, which suggests that personality has three independent dimensions: neurotic-stable, introvert-extrovert and tough-tender minded. Your personality can be completely defined, according to this inventory, by working out where you fit along each of the three dimensions.

This scale was undermined by London neuropsychologist Jeffrey Gray, who pointed out that anti-anxiety drugs like alcohol or the benzodiazepines simultaneously reduce scores for neurosis and introversion. If a single drug can affect two axes at once, it suggests that they are not independent.

More recently, Bob Cloninger at Washington University, St Louis, focused on temperamental factors underlying chronic anxiety and was at pains to draw together all strands of research. He felt that psychologists like Eysenck had paid too little attention to the brain's neuroanatomy, information-processing systems and environmental factors. Equally, neurobiologists and learning specialists sometimes neglected the multi-dimensional structure of personality. He synthesised a mass of information on anxiety, including studies on the family, the structure of personality, neuropharmacology, neuroanatomy, behavioural conditioning and learning in both humans and animals. Out of this, he came up with a different three dimensions of personality.

Cloninger called his dimensions novelty seeking, harm avoidance and reward dependence. They are independent of each other and are inherited separately with different genes responsible for each. He predicts that different chemical messengers in the brain mediate each of the dimensions: dopamine for novelty seeking, serotonin for harm avoidance and norepinephrine for reward dependence. Scores on these dimensions describe personalities and can even predict responses to different situations, he says.

For example, harm avoiders tend to be apprehensive, cautious, shy and easily fatigable while their opposites are fearless risk takers. Novelty seekers are impulsive, yet intolerant of pain while their opposites are rigid, stoical and reflective. Reward dependents are sentimental and moody while their opposites are detached, perhaps socially isolated.

Anxiety disorders are related most closely to harm avoidance. General anxiety, panic disorder, agoraphobia, social phobia and obsessive-compulsive disorder are all associated with high harm avoidance. Novelty seeking among phobics is similar to that in the general population, and reward dependence varies between groups. Social phobics are low in reward dependence, for example, where obsessive-compulsives may be high. But the strongest link between anxiety and temperament lies in high harm avoidance.

Cloninger arrived at this picture of personality after a careful analysis of the literature and the next step was to back it up with practical work. In the laboratory, he found that harm avoiders had a more intense response to signals of punishment, or of not getting a reward. They were generally slower and more cautious than others. When the test included an unpleasant stimulus or distracting signal, such as in the Stroop test described in the previous chapter, they dealt with it more slowly than those with uninhibited, spontaneous reactions. Harm avoiders are more easily distracted or delayed.

In the clinic, this translates into shyness, fatigability, fearfulness and anger directed internally rather than towards the world. All of which provides fertile soil for the growth of anxiety and mood disorders. In life, harm avoiders are cautious and inhibited, perhaps taking longer than usual to make decisions.

When Cloninger looked at large groups of people over a

period of time, he found that women overall scored significantly higher on harm avoidance than men. This ties in with
women's increased vulnerability to anxiety disorders. Women
were also lower in novelty seeking and generally more reward
dependent or socially sensitive.

Harm avoidance changes little with age, and reward dependence decreases slightly but not much. Novelty seeking
decreases significantly as people get older, which you might
expect. We are not surprised when eighteen-year-olds want
to go hang-gliding or bungee jumping; we are rather more
taken aback if eighty-year-olds have the same ambition.

However, harm avoidance can change radically, soaring in
a period of acute anxiety or depression. Novelty seeking and
reward dependence, by contrast, tend to remain quite stable.

Harm avoidance, then, is higher in women than men,
generally stable throughout life but liable to increase if someone is acutely anxious. Thus far could perhaps have been
predicted, but Cloninger's work revealed some unexpected
relationships. Criminality and anxiety are not usually considered in the same breath, but he thought they might be
related and was backed up by a Swedish study. Researchers
followed 807 men who were adopted at between four and
eight months. They found that if the biological mother had
been involved in crime, the adopted son was likely to have
low levels of anxiety. What they may have shared, Cloninger
says, was the trait of low harm avoidance. The mother had
a lack of concern for the law or the consequences of being
caught. The son was brought up in a different environment,
but took with him a biological tendency not to worry and
became almost immune to anxiety disorders. This study alone
is insufficient for us to start re-writing theories of criminality,
but it hints that low harm avoidance could be at least partly

inherited. These men could not have learnt their carefree attitude from their natural parents because they had been adopted as babies.

Reactions to anger could also be predicted by the dimensions. Harm avoiders are as likely as anyone else to feel resentful and angry but they tend to contain their feelings. This may be convenient for colleagues and aquaintances but could feed a low self-esteem. On the other hand, novelty seekers readily express anger, which may protect their self-esteem but not their reputation in society if they get into regular fights.

Cloninger and his colleagues assessed the personalities of 4,000 adult twins, some identical, some not. Computer analysis of their results was encouraging, especially among men. It showed that harm avoidance was independent of other dimensions explored, including novelty seeking and reward dependence, but also persistence and others. Cloninger became convinced from this that the basic genetically controlled traits behind personality are indeed harm avoidance, novelty-seeking and reward dependence. Others, however, have taken a slighly different tack in looking at temperament in relation to anxiety disorders.

## The Inner Child

Take a four-month-old baby, swing a brightly coloured mobile in front of his face, burst a balloon behind him, and what do you get? Any one of a variety of reactions, probably, from crying and fretting, furious kicking of legs and pumping of arms to very little response at all. But according to researchers at Harvard, Massachusetts, you also get a hint as to how this child will behave on his first day at school.

Psychologist Jerome Kagan specialises in children's temperaments and his recent work suggests that aspects of temperament may become obvious long before a baby can walk. Some of Kagan's ideas have been discussed in earlier chapters and, like Cloninger, he is a firm proponent of the influence of genes and the idea of in-built traits.

Kagan's interest in temperament started in the late 1950s when he was working in Ohio. An on-going study set up in 1929 included adults who were first interviewed when they were three years old. Kagan and his colleagues visited them at home to compare their adult personalities with the notes made years previously. They found that little of the child remained in most of the adults. However, one quality seemed to be preserved over a lifetime. Children who were fearful of the strange and the new grew into adults with the same reservations. Kagan and his colleagues called this passivity to threat; it is now known as inhibition.

Twenty years later at Harvard, Kagan was studying the effects of day care on babies. Half the infants were Chinese American, half were European Americans, and the difference between the groups was striking. The Chinese American babies as a group were quiet and fearful and the European Americans were noisy and demanding. The children were so young that Kagan concluded that it had to be caused by in-built differences in the babies' temperaments.

Remembering his previous work in Ohio, Kagan went on to study inhibition. It is prevalent and easily observed, a help to researchers, but may also be relatively permanent. He set up a series of long-term and ultimately fruitful projects among a privileged group of infants. Their mothers were all white and middle class and had textbook pregnancies. They neither drank alcohol nor smoked; anyone who even had too much coffee was excluded. There were no problems at birth, no premature infants and no abnormalities in the children within this group. Kagan wanted to look only at temperament and aimed, as far as possible, to remove complicating factors within children's environments from the study.

The studies started with two groups of toddlers put in slightly strange situations. For example, twenty-one-month-old infants were encouraged to carry out simple tests with an unfamiliar woman, to approach and touch a robot, and then left alone in the room. Thirty-one-month-old infants were observed playing with an unfamiliar child of the same age and reacting to a woman with a plastic cover over her head and top half.

Some children were markedly more fearful than others, fretting or crying, staying close to their mothers and refusing to approach unfamiliar women or robots. They were classed as inhibited. Others showed no hesitation or distress and were considered uninhibited. The remainder displayed inconsistent behaviour and were not followed up afterwards.

The two groups of children were followed up at regular intervals until they were thirteen years old, with tests altered to make them appropriate for their age. Not all of the children maintained consistent behaviour in the ensuing tests but some did. Overall, a third of the children described as inhibited or uninhibited at twenty-one or thirty-one months were classed in the same way at seven years old. It was rare for children to change from one category to the other and only three children from the two samples did so.

By seven, inhibited children typically stay to one side of a group. They might read or draw or just stand and watch other children playing. They are subdued, shy and withdrawn. They are not necessarily unpopular but they avoid boisterous play and spend a lot more time alone than most children. Uninhibited children are quite the opposite. They smile and laugh readily and are full of vitality. They are confident and happy to go up to other children to talk or play.

The Harvard children changed little as they became adolescents and at age thirteen most had the same demeanour they had shown at seven years old. The uninhibited group smiled on meeting an unfamiliar examiner and often smiled if they failed a difficult test, as if laughing at themselves. They chatted easily and volunteered conversation with a psychiatrist. The inhibited group were quieter, smiled much less often and answered questions put to them without elaboration.

It seemed that an hour in the laboratory with a toddler was often sufficient to uncover aspects of their character that were still intact eleven years later. The inhibited and uninhibited temperament was a more or less ingrained part of these young people, right through their childhood and adolescence, and it led Kagan to wonder if it might be possible to predict temperaments from an even earlier age.

Doreen Arcus, a colleague of Kagan's, put a group of 500 four-month-old babies through a variety of stimulating tests. Bright mobiles were swung in front of their faces and cotton swabs dipped in dilute alcohol put under their nostrils. Some moved around vigorously in response but were not upset. Others were disturbed but remained relatively still. These groups she called, respectively, aroused and distressed. Other so-called high reactive babies kicked their legs, pumped their arms and cried or fretted. Low reactive babies moved only occasionally, rarely fretted and were more likely to smile sometimes.

Babies were assessed again over the ensuing years. At four

and a half, the children who had been classed as high reactive at four months smiled less, were less likely to make spontaneous remarks and were more intimidated by the examiner than those who had been classed as low reactive. In one test the examiner took out a photo album, told the child which was her favourite photo and asked him or her to tear it up. High reactive children seemed afraid to disobey the request and typically tore a small corner of the photo. Low reactive children were more likely to ask why they should do it, or just refuse. One child simply said, 'Oh-oh, no way, I'm not doing this,' closed the book and walked away. The children who refused the request did not seem perturbed by the incident.

This work suggests that simple laboratory tests on four-month-old children can predict their behaviour as children and even as young adults. The prediction is far from being 100 per cent accurate, but Kagan's team believe that a true inhibited temperament is partly in-built or genetic. Elements of the temperament are programmed into the child's physiology and make the child much more likely to develop into a shy, withdrawn, undemonstrative adult.

This should not be taken to mean that all babies with a highly reactive disposition inevitably become inhibited children. Many children find a way out. Our behaviour and personalities develop as they do with contributions from all areas of life. Interwoven with physiology is experience and inhibited behaviour can be learned or unlearned by life's events. Some children lose the bulk of their inhibited behaviour and a rare few go on to appear quite uninhibited. Equally, some low reactive babies become inhibited after an unfortunate or deeply traumatic experience. One girl who had been uninhibited when very young became inhibited after her father committed suicide. The tragedy altered her personality altogether and she behaved afterwards like the children who were born with an inhibited temperament.

However, the truly inhibited child has a physiology that is quite distinct from that of a child who has become shy and withdrawn through experience, according to Kagan, even though the behaviour of the latter may be just as difficult to alter. He believes biological features of an inhibited temperament may exist even before birth. His colleague Nancy Snidman measured maternal and foetal heart rates in another group before birth and found that babies who became high reactive were likely to have had heart rates of more than 140 beats per minute in the womb. The boy with the highest heart rate in utero, of 170 beats per minute, became a highly fearful and nervous child. The girl with the lowest, at 120 beats per minute, became a relaxed and not easily distressed child. After birth, differences in heart rate were too inconclusive to predict later behaviour. But, in general, infants eventually categorised high reactive tended to have had higher heart rates as babies.

Present knowledge will not allow observation of an unborn baby to predict how he or she will behave as a child or young adult. But Kagan believes that in future more sensitive or specific tests may differentiate inhibited and uninhibited children. Differences may exist in the womb, he says, but heart rate measures are just not sensitive enough to pick them up.

This is not as unlikely as it sounds if an important component of these temperaments is programmed into us through our genes. Kagan believes that genes involved in temperament may exert their influence via the sympathetic nervous system. Heart rates give an indication of the activity of the sympathetic nervous system and there is no reason to suppose that this only starts at birth. If genes can influence our physiology as children and adults, then it makes sense to think they can influence the physiology of an unborn child.

The involvement of the sympathetic nervous system has already been discussed in earlier chapters but, essentially, high reactive infants may have a system that responds more readily to stimuli. The ancient, limbic sites of the brain which are responsible for primitive emotions and reactions like fear are linked to sympathetic sites in the cardiovascular system. According to Kagan, these links are more easily excited in inhibited children than others. Heart rates are only indirectly related to children's reactions to new people, places and situations but they give an indication of what the child is experiencing.

Other tests on inhibited children aged five and seven found that in strange and new situations they had larger pupil sizes, greater acceleration in heart rates and larger changes in blood pressure than other children. This again suggests that their sympathetic nervous systems are more active.

An overall pattern started to emerge. High reactive children have high heart rates at four months old, appear fearful and smile less in the second year, and are more likely to become inhibited children and adults. Low reactive children have low heart rates and fear scores and are frequent smilers. They become exuberant children, full of vitality and are likely to grow up to be extroverts.

So far at least, not all of these features have been found in all studies. Some results have been found in one sex and not the other, some have been found in inhibited but not uninhibited children. Behaviour, emotion and physiology move as one for some people, some of the time, but by no means always. Shy and withdrawn children who are feeling fearful may have a wildly accelerating heart rate, but then again they may not. The research is in its early stages and many loose ends remain to be tied.

Arcus believes that inhibited and uninhibited infants inherit two independent qualities: the ease of arousal and the mood that arousal produces. Some children smile when stimulated, others become distressed. In her group of 500 babies, those she termed aroused may have had a physiology as easily triggered as the high reactive children. They moved around vigorously during the laboratory tests, but they did not become upset. They had only one of the two components. Other 'distressed' children were upset by the tests, but did not kick, wriggle or appear particularly stimulated. They, too, had only one of the two components.

High reactive children are easily aroused and usually go on to become distressed. Low reactive children are not easily aroused, but when they are, tend to become happy or joyful.

Kagan's group believes that these qualities are partly determined by genes. But genes cannot be the full story. Even in the womb, different foetuses have different experiences. French researchers found that if they stressed female rats in the last week of their pregnancy, their rat pups bore a psychological scar after birth. The scientists removed the adrenal glands, which produce the stress hormone corticosterone, from half the pregnant rats and gave them an artificially constant level. The others were left intact and all were put under an unpleasant bright light a few times. Once the pups were born they were given the same experience. The researchers found that rat pups born to the intact mothers took far longer to recover from the experience than those whose mothers had no adrenal glands. The rats whose mothers had normal adrenal glands had presumably experienced a surge of corticosterone while their mothers were stressed. This could have desensitised receptors in the pups' brains permanently and left these rats less able to switch off their own production of the stress hormone.

If it were possible to extrapolate from rats to people with any confidence, this finding would imply that a stress-free gestation gives a good start in life. What may be more pertinent, though, is the finding that even environmental influences can start before birth.

## Family Likeness

Genes may influence our destiny, but they do not control it. More than a third of Arcus's high reactive infants were not fearful in their second year and a few were even fearless. This interested her, since it suggested that parenting style could be halting the expected development of the inhibited temperament. She visited the homes of high and low reactive children and watched how mothers related to their children.

Arcus found that nurturing parents who consistently protected their children from minor stresses made it more rather than less likely that they would become fearful and timid as they grew older. Children seemed to benefit from having to keep a few house rules. High reactive children were helped by mothers who set firm and direct limits on their behaviour. The children were all living in comfortable and supportive homes and there was no question of neglect or abuse. Simply taking something away from a baby and saying, 'No, you can't have that', creates stress, but if the child learns to deal with such stressors within a nurturing and caring environment, this seems to protect against developing fearful behaviour later.

Research with animals tentatively supports this finding. Baby rats who were separated from their mothers for five minutes a day later performed better in tests that required them to cope with some stress than rats that had never been separated. The brief separation appeared to equip them to deal with mildly unpleasant events. However, rats which were separated from their mothers for three hours a day could not cope at all. They lacked sufficient support at a crucial stage in development and became more rather than less anxious as a result. Even rat mothers have to strike a balance between over-protectiveness and neglect.

The home environment may also influence the sex ratio of different temperaments. Kagan found that baby boys and girls are equally likely to be high reactive, but girls are more likely to translate this physiology into inhibited behaviour than boys. This is probably cultural in that virtually every known society has praised male bravery, and forgiven or even applauded shy, submissive behaviour in girls and women. An inhibited boy seems more out of the ordinary than an inhibited girl and parents are likely to go to far greater lengths to encourage and promote more outgoing behaviour.

Arcus suggested that women may come to motherhood with firm views on parenting styles, or develop them early on. One view is that life should revolve around the baby and all demands should be met. Another is that the baby has to fit into the home life. He or she will be cared for, but will gradually have to learn what is or is not acceptable. Arcus suggests that for inhibited children the latter approach may be most helpful.

Research in Cambridge, UK, points to a more complex picture. Dr Lynne Murray has suggested that a mother's influence over her baby may be overstated. Current baby-care literature emphasises the importance of the mother's behaviour in virtually everything the baby does, from meeting development targets to laying foundations for future emotional security and intellectual development. Murray has demonstrated, equally, that babies influence their mothers. Irritable babies who were easily upset and then difficult to soothe made up 17 per cent of the study group she followed. Having an irritable baby tripled the mother's risk of being depressed six weeks after the birth. Murray believes that many women with difficult babies have an implicit sense that it is their fault, and she thinks a more realistic approach to baby care, which recognises that even babies are individuals, would help.

Murray agrees that babies are inherently different, right from birth. This is in line with evidence for the genetic component of temperament and is strengthened by research into the behaviour of other species. As mentioned in chapter 3, rats and other animals can be bred to become timid. Uhde's work with nervous pointer dogs, discussed in chapter 4, demonstrated that animals kept under identical conditions had different temperaments, again because of their genetics. Even without selective breeding, about one in five rhesus monkeys is fearful and timid in new surroundings. This percentage is similar to the numbers of children found to be inhibited in Kagan's laboratories.

If a relatively consistent percentage of both animals and humans are inhibited, it suggests that the quality has been conserved throughout evolution and might therefore carry a survival advantage. Or at least have no disadvantage. In ancient times, it is easy to imagine that hunters who were timid and fearful might have taken fewer risks than their bold friends. They may have been less successful at bringing home the bacon but they might also have been more likely to return home at all.

Today, Western societies value gregarious, open, friendly people and it seems that the uninhibited have a head start. But in the long term there are few drawbacks to being inhibited. Inhibited adults may be unlikely to forge a career as stockbroker or criminal barrister, but there are plenty of prestigious, well-paid alternatives. Artists, scientists, academics and computer programmers can have highly successful careers without being forced to demonstrate advanced social skills. Inhibited people may choose different but no less favourable paths through life.

## Inhibition, Fear and Phobias

Or at least most do. While Kagan was studying the temperaments of privileged, healthy children in Harvard's psychology department, Jerrold Rosenbaum was working at the general hospital among adults with panic disorder and agoraphobia. These adults had often been anxious as children, suffering particularly from separation anxiety and reluctance to leave a parent, typically their mother, to go to school. More than half of all women with agoraphobia had separation anxiety as a child.

Other studies had shown that children whose parents had agoraphobia had higher than usual rates of separation anxiety themselves. This all tied in with the well-established observation that panic runs in families and that children living with a parent who has a psychiatric disorder are themselves more likely to develop problems.

Kagan's descriptions of inhibited children sounded familiar to Rosenbaum. They were reminiscent of what his patients told him about their behaviour as children. Stories were commonplace of withdrawn behaviour, fear of anything new, difficulty getting to know other children and a horror of being singled out in the classroom. Rosenbaum started to wonder whether the inhibited behaviour described by Kagan among healthy children was putting them at increased risk of developing anxiety disorders later on in life – whether the inhibited behaviour meant more than a shy personality and was actually a risk factor for agoraphobia and social phobia.

The first collaborative study between Rosenbaum and Kagan was a rather haphazard affair. Rosenbaum gathered a group of children at risk of developing anxiety disorders because of their family history and sent them for analysis in Kagan's lab. The children were a mixed bag. Some of their parents had panic disorder and agoraphobia, some had other psychiatric disorders such as major depression. They were compared with Kagan's children, who were drawn from a population with no obvious problems at all.

The results were striking. Rosenbaum's children at risk, selected on the basis of their parents' problems alone, had telltale signs in their own behaviour. Where parents had panic disorder and agoraphobia, their children were more likely to be inhibited than children whose parents were healthy.

Furthermore, inhibited children, whatever their parents' health, were more likely to have anxiety disorders. Relatives of Kagan's inhibited children – parents, brothers and sisters – were more likely to have anxiety disorders than relatives of uninhibited children. The parents were more likely to be avoidant, anxious and socially phobic, and to have had anxiety disorders as children.

Inhibition foretold later problems. Children classed as inhibited in one assessment were more likely to have an anxiety disorder when followed up three years later. Some children were classed as inhibited at some assessments and appeared open and relaxed at others. Some, though, were inhibited at every assessment, and these children were called 'stably inhibited'.

Kagan and Rosenbaum re-examined their findings and discovered that the increased risk of anxiety disorders was almost entirely accounted for by the group that was stably inhibited. These were the children with the anxiety 'bug' and were the ones at real risk.

Rosenbaum believes that inhibition in a child might be especially potent as a sign of future trouble where that child is already at risk of developing anxiety disorders. If parents are healthy and sociable, a child's inhibited behaviour might add little to their risk. It might just mean that they are less likely to choose a career in sales or at the bar. If a parent has agoraphobia, though, the child's inhibition may mark them out at an early age as being likely to suffer from anxiety.

These children may have a constitutional predisposition for developing anxiety and a life-long risk of problems. They may have a marker for the so-called anxiety diathesis, which describes those who start out as reactive, colicky, sleepless and difficult infants. They develop into inhibited toddlers and stay on the fringes of play in school. As adolescents they have a range of social fears and typically go on to have panic attacks at an important stage of life such as when they are leaving home for college – rather earlier than other people develop panic disorder.

Rosenbaum believes that the diathesis, or lifetime predisposition for anxiety disorder, explains why people so often have more than one diagnosis. Panic is surprisingly common among children and Rosenbaum's colleague, Joe Biedermann, collected data on 700 children referred to the clinic with panic. They described signs and symptoms relevant to so many different disorders that they could be given an average of seven different diagnoses. Rosenbaum says the idea that they have seven different disorders is incredible. It is much more likely that they have a general disposition for anxiety that can manifest itself in different ways. Their panic disorder, social phobia, obsessive-compulsive symptoms and possibly depression are linked at the heart of this diathesis. Over time, given different life experiences, their predisposition expresses itself in different ways to produce different disorders. So an anxious child brought up by an agoraphobic parent is at risk of developing agoraphobia themselves. But if the parent has social phobia, a child with exactly the same predisposition has an increased risk of developing that, and not agoraphobia.

Rosenbaum was interviewed for a television item on agoraphobia and panic alongside a mother with agoraphobia whose ten-year-old child appeared to have panic disorder. The mother was asked when she first noticed the signs in her child. She replied that at four months old a mobile over the baby's cot over-stimulated and upset the baby so badly it had to be taken down. Friendly adults who rattled keys to amuse her were greeted with screams, tears and a squirming, kicking, obviously unhappy baby. This little girl was exceptionally fearful and timid and always had been.

She is a good example of a child at risk. Her mother has agoraphobia, she was highly reactive as a baby and the panic started young. She has many of the ingredients that can lead to a lifetime of anxiety disorders. She is at increased risk of agoraphobia, social phobia and probably more likely to develop situational and simple phobias as well as obsessive-compulsive symptoms and possibly depression. Not that she will go through life with all of these problems, but difficult experiences and events are more likely to precipitate problems for her than for other people. Her saving grace may lie in the recognition of her plight. Careful intervention, as discussed later in this chapter, might reduce the risk of her sliding into one of a number of anxiety disorders.

Rosenbaum's diathesis is rather like the British idea of neuroticism which puts people at risk of developing any of a range of anxiety disorders or depression. It is out of step with the typically American concept that different disorders arise out of different biological risk factors and are passed down from parent to child quite separately. Kagan believes rather that a child's inhibited temperament may increase the later risk of developing social phobia only. Agoraphobia is not linked to an inhibited temperament, he says.

It is tempting to speculate that this difference between the two collaborators is at least partially explained by their background. Kagan, the academic, studies ideas and it makes more sense from a theoretical perspective for inhibition to be linked with social phobia. Rosenbaum's work takes place in the clinic and he is surrounded by people with a mixture of anxiety disorders, who may be difficult to categorise so neatly.

Rosenbaum's views also contradict theories discussed earlier, notably the work on respiration by the Columbia group. Children who have panic attacks have respiratory symptoms, like adults. However, the Columbia group assumes that panic disorder is a disorder of respiratory control. A defect leads to respiratory attacks, or panic attacks, and then complications such as agoraphobia develop. The panic attack comes out of the blue and the problems follow. Rosenbaum, however, insists the fear behaviours were there long before the first panic attack. People may become worse and develop new symptoms after the attack but many already had phobias and showed other signs of fear before the first hint of respiratory involvement.

Someone with the physiology that underpins the anxiety diathesis may live with constant but not crippling anxiety for years. It is possible that over time their condition erodes their resistance to respiratory problems so that eventually their threshold for panic is lowered. One day, for no very obvious reason, they have their first 'attack'. This is Rosenbaum's way of reconciling the two strands of research but it is conjecture and the central tenant of the two theories remains inconsistent. The Columbia group believes the attacks arise with no warning; Rosenbaum's team believe people live all their lives at increased risk of anxiety disorders.

Another explanation is that two separate populations develop illness in quite different ways. One group really does have a couple of respiratory attacks with no warning and then develop complications. The others have similar spells following a lifetime of low-level anxiety, which becomes much more intense afterwards. In this way, both theories could be true when applied to the appropriate people.

One indicator that a child has the anxiety diathesis could be sleep panic attacks. Discussed in chapter 4, these were described by Uhde's team as occurring specifically in adults. Rosenbaum, however, says that children at risk describe dream anxieties and night terrors. He believes it is likely they are having sleep panic attacks which are reported imprecisely

by clinicians and researchers. Like Uhde, he believes that
sleep panic attacks are spontaneous physiological events, not
triggered by learned cues. People are woken from sleep without any environmental or cognitive prompt, which may mean
in turn that those who have had sleep panic attacks are more
likely to have the anxiety 'bug'.

Investigating this further, Rosenbaum's team divided a
group of anxiety patients at the clinic into those who had
experienced sleep panic, and those who had not. They found
that a history of sleep panic was linked with higher rates of
other anxiety disorders. This group were more likely to have
generalised anxiety, social phobia and depression. They also
tended to have been ill for longer and were more likely to
have a history of childhood anxiety disorder. He concluded
that sleep panic in adults might be a distinguishing marker
for the anxiety diathesis.

Rosenbaum, Kagan and Cloninger have large areas of overlap. Kagan's behavioural inhibition, for example, might be
easy to spot in the clinic, but Cloninger believes it is a com-
posite. Inhibition is linked with high harm avoidance, he
says, but probably also with low social warmth, that is, low
reward dependence and a lack of exploratory pursuit or low
novelty seeking.

Rosenbaum's more general neuroticism is a combination
of high harm avoidance, again, but also high novelty-seeking
and high reward dependence. So, according to Cloninger,
neurotics have inherited three quite separate dimensions.
Rosenbaum is more inclined to think that people lie along
a continuum for anxiety. At the point where the anxiety is
so great that it starts to interfere with life, it is classed as a
disorder and it is time for physicians to become involved.

Kagan believes that the genetics behind the inhibited tem-

perament are inherited in a lump. The inhibited temperament, he says, is a discrete category, as different from the
normal healthy population as, for example, diabetes. People's
blood sugar levels may lie on a continuum, as may their
insulin production, but some have a healthy pancreas and
others do not. In the same way, anxiety levels lie on a continuum, but people either have or do not have an inhibited
temperament, claims Kagan. It may be at the heart of anxiety
for some, but it does not come in degrees. Cloninger, by
contrast, believes that genes exist not for the inhibited temperament, but for dimensions which contribute towards it.

Differences between Cloninger, Kagan and Rosenbaum
may exist more in theory than in the clinic. Cloninger's
description of an anxious temperament sounds uncannily
like Rosenbaum's anxiety diathesis.

Cloninger charted the development, from birth to adult-
hood, of 300 men in California. They and their relatives were
questioned. Underlying temperament, intelligence, family
environment and a variety of physical characteristics were
considered.

Normal development and social maturation is thought to
come in fairly definite steps. Babies are pre-social, two-year-
olds impulsive. By the time a child is three or four they are
opportunistic, by six or so they conform to reason most of
the time at school and once they are at high school they are
becoming self-directed and conscientious. From there they
move on to becoming good team players and eventually
mature as older teenagers.

Cloninger says that those with an anxious temperament
tend to move early to the conforming stage, but then get
stuck. They do as they are told but are unlikely to act on
their own initiative. They have a difficult time becoming

self-directed, self-confident and setting out clear-cut goals in life.

Like Kagan, Cloninger recognises the importance of environment and says that this tendency can be compensated for by the style of parenting. Men who were high in harm avoidance but raised in a supportive home environment, which was not over-protective and encouraged autonomous decision-making, moved up to a more conscientious level. So Cloninger's view, in practice, is compatible with Kagan's. Both have a hefty genetic contribution but this interacts with home environment and, says Cloninger, intelligence.

One difference is that Kagan studied children while Cloninger concentrated on adults. Children have less opportunity and are less able to make decisions about how to cope with their own temperament. There may be far fewer real categories of children than of adults. A group of people who start off with an identical inhibited temperament will, by the time they grow up, each have developed their own way of dealing with it. Categories that are obvious and common in adults may simply not exist in children.

Rosenbaum, Kagan and Cloninger all agree that temperament is an enduring quality which can change little throughout life. Some temperaments might increase the risk of anxiety disorders and doctors and parents may become better able to single out children at risk from an early age. But the end point of this research must be to find ways of reducing that same risk.

## Triple Jeopardy

Children born to anxious parents could have triple jeopardy, Rosenbaum says. They may have inherited an anxious tendency, have a parent who is modelling anxious and inhibited behaviour and they may be receiving sub-optimal parenting if the parent is sick, absent or unable to care.

This was the situation for Elizabeth's daughter, Anne. Elizabeth was struggling to deal with her own phobia, and with bringing up a child alone. 'The poor child didn't have a very good start,' Elizabeth says now. 'My husband left us very suddenly. I didn't enjoy bringing up my daughter because I was grieving for my husband.' In addition, she believes that she and her daughter have such similar personalities that they must be partly shaped by genes.

Anne was diagnosed only recently and is hoping that her current treatment, of drugs and counselling, will turn out to be short term. For Elizabeth, long-term medication has been necessary. She tried behaviour therapy with a counsellor whose goal was to take her into a hospital sluice. They started with a video of someone being sick, and she took it home to practise. 'I was supposed to watch it every day, gradually turning up the sound. But I couldn't cope with it. It was a waste of time. It seemed that once I started on the course, I was concentrating on the problem all the time and I got worse.' Counselling that delved into her childhood got nowhere. 'I couldn't remember anything particularly important,' she says.

Elizabeth has taken anti-anxiety drugs continually since 1965. She came off them while pregnant, and had one attempt to stop but the withdrawal was so terrible that she felt it wasn't worth it. She has been much improved for the past four or five years, since she started on Prozac. She is more relaxed about housework, more laid back generally, and able to eat out on occasion.

The phobia remains but has less hold on her life, at least for the time being. But she is dreading the arrival of grand-children. 'My biggest worry is my daughter having children. They are going to want to stay the night and I will have to babysit. I don't know what my daughter's plans are but I think grandchildren are at least two or three years away. I try not to think about it.'

## Challenging Fate

Work on personality and temperament is fascinating in theory, but the next stage must be to determine ways of helping a child at risk. Can parents, doctors and other pro-fessionals intervene early and stop anxiety snowballing through someone's life?

Rosenbaum believes that among children whose parents have panic and agoraphobia or social phobia, the child who is inhibited is most at risk. And they may need help.

The least controversial intervention is energetic treatment of parents. At a recent meeting of the American Psychiatric Association, Rosenbaum urged his audience to remember that they are treating a system, not just one person. If the parent does better, it may prevent severe pathology in the child. He called on professionals to use drug treatments, cognitive and behavioural therapies, family therapies. 'Do it all!' he said.

Professionals need to be aware of what the inhibited child

is experiencing and to encourage parents to recognise it as well. In a way, the parent needs to become the child's therapist. The inhibited child may have exaggerated physical responses to minor stimuli – things that would make another child laugh may give them a fright. They may frequently be unpleasantly aroused with an uncomfortably high heart rate, dry mouth, dizziness, even feeling frozen with fear at quite unremarkable events. When something really upsetting happens, perhaps being chased by a big dog or stung by a swarm of bees, the inhibited child may be the one most likely to develop a phobia as a result.

Children learn ways of thinking and behaving from their parents and an inhibited child with an anxious parent is more likely to see and imitate social or emotional disabilities. But parents' intervention can help if they are aware that their child feels differently and needs more reassurance than other children. New and challenging situations are that bit harder for the inhibited child to face. Starting school, a milestone in any child's life, is likely to be especially stressful and the inhibited child may need more preparation, rehearsal and explanation than others. Parents can use exposure therapy principles and perhaps take the child down to the school, several times if necessary, to meet teachers in advance. It's a good idea for any child, but this child may need it most of all.

Rosenbaum and his colleagues have put together a manual to help parents intervene. They will be trying it out with parents with panic disorder and hope to show that it helps their children. If it does, they will give it to parents who are not themselves anxious but who have inhibited children.

The next, most controversial, step is for doctors and other professionals to treat the child directly. This is not

appropriate for inhibited behaviour alone, but where the
child is showing signs of anxiety, it may be helpful. There is
always a reluctance to put children on drugs, especially for
a long-term, ingrained condition, and treatment would typically start, says Rosenbaum, with cognitive and behaviour
therapy. But he says that children's development may be
influenced by their physiology. They over-react to normal
stressors with so much intensity that they become distressed.
They learn to avoid situations and to develop the negative
ways of thinking associated with anxiety disorders. Rosenbaum argues that if the physiological component of their
reaction could be turned down, so that they did not have
the unpleasant beating heart, or dizziness, then it could be
as effective as a behavioural therapy in altering children's
lives. Anti-anxiety drugs include treatments to tone down
the noradrenergic system, the selective serotonin re-uptake
inhibitors or benzodiazepines.

Giving a child a drug therapy over a prolonged period *is*
controversial and opposed vehemently by some parents but
Rosenbaum is firmly convinced that children with a clutch
of risk factors may sometimes be best served by it. He says
the effects are usually dramatic and from one week to the
next children are transformed. They start enjoying life, make
friends, seem more confident, and parents typically wonder
why they waited so long before trying a drug treatment. The
important question is whether children who grow up feeling
comfortable and having friends in the end have a reduced
chance of developing an anxiety disorder later in life.
Research is difficult as few children are entered into clinical
trials and it takes many years to get an answer.

Europeans tend to be more reluctant than Americans to
give children drugs. Ritalin for hyperactivity is used in

Europe, but many more American children receive it. An anxious or inhibited child's problems are less pronounced than those of a hyperactive child and it is hard to imagine many European parents agreeing to this approach. However, Rosenbaum is insistent. Drugs may be used for short periods to get someone through exams or job interviews but he believes the anxiety diathesis is a lifelong condition which may need lifelong treatment. Just as a young person with diabetes can expect to need treatment all their lives to control their metabolism, someone with the anxiety diathesis may have to accept that they will need treatment over a long period of time to regulate a system which cannot regulate itself.

This view sounds rather pessimistic compared with much of the other work in this book, but most people with phobias, especially the specific phobias, will not have the anxiety diathesis. Having it increases the chances of having phobias, but only a fraction of phobics have it. And for many of those that *do* have the diathesis, receiving a realistic assessment of what happens to them and an accurate description of how they feel is very reassuring.

The good news is that, however strong the in-built or genetically driven temperament, research is showing that the environment has a powerful influence. Parents who have struggled with anxiety all their lives do not wish to pass it on to their children and this work is starting to provide hints about the different ways in which children at risk can be steered away from a path strewn with anxiety disorder.

Having children can be a turning point, an opportunity to change things. Diana and Kay, whose experiences were described in chapter 5, and Nicola, in chapter 3, were all prompted to address their own phobias by the thought of their children picking them up. More than anything else, they wanted their children to be free from their own disorders, agoraphobia, dental phobia and social phobia.

All three conquered their phobias, and cut out two of Rosenbaum's three strands of jeopardy. They ceased to be role models for anxiety and their parenting improved at a stroke. The remaining strand, the child's in-built temperament, may be addressed by long-term drug treatment. Alternatively, just an increased awareness of what the child is going through can help, with extra rehearsal and explanation before a big event. The findings on parental style are preliminary but suggestions that over-protection is unhelpful and that children benefit from firm limits on their behaviour are interesting.

Perhaps the most important message from this work is that help is available now, and that ongoing research is set to improve it. The more we understand why people become vulnerable to anxiety disorders, the better the advice will become. Research into temperament and personality may eventually show us how to ward off the agony of anxiety and phobias before they are ever able to take hold.

## CHAPTER 8

# Gender and Hormones

## A Pregnant Pause

Some women moan about pregnancy, but for Alison it was a dream. It meant she could go out without fear. Shopping became a pleasure and she started to see friends she had stopped contacting. She bloomed, physically, mentally and socially.

Her first panic attack had occurred two years previously while she was shopping in a department store. She was browsing around the perfume counters when she became aware that her heart was beating wildly. She started to shake, she felt sick, she felt as if she could not breathe. She had an overwhelming sense that she was out of control. It was a terrifying experience.

Initially, she assumed it was a one-off, but within a few months she was having attacks like this every day. She began to avoid people and shopping malls, and her social life became increasingly restricted until her pregnancy unexpectedly gave her the old life back again. Unfortunately, the moment her baby was born, the old feelings came back.

This time, it was worse than ever and she felt unable even to drive very far, a major restriction when she was looking after a young baby. Her panic attacks and avoidance continued for three years. Then, when she became pregnant again, she was instantaneously better. She felt she had more control over her life and could drive freely without fear.

After her second child was born, her panic returned again and was so acute she was prescribed medication, which helped but did not eliminate her fear. Her only true respite came during the nine months of her third pregnancy.

This story is interesting and not altogether unusual. It is based on one of three cases reported in the prestigious *American Journal of Psychiatry*. Researchers led by David George described three women with long-standing panic disorder and agoraphobia. Pregnancy coincided with dramatic improvements for all of the women until or shortly after delivery. All were pregnant more than once and they had the same pattern of improvement during each pregnancy, followed by a rapid return of panic and phobias afterwards.

One of the other women was able to cope with crowds while pregnant and had a new-found confidence that everything would work out. She felt happier than usual, closer to friends and family and had higher self-esteem. Perhaps most poignant of all was the story of the third woman, who had extreme phobias of height, driving a car and eating food out of cans. While pregnant, she could drive long distances from home and look out of the windows of tall buildings without fear. Her panic attacks were dramatically reduced. Years later, she remarked that in terms of her emotional well-being, her pregnancies were the best periods of her life.

Despite cases like these, investigation into the impact of female hormones on panic and anxiety has only just started and early research has so far failed to find a conclusive link between specific hormones and specific effects.

The importance of gender in phobias is not in dispute.

Agoraphobia is diagnosed about three times as often in women as in men. Specific phobias are also much more common. Women make up between 75 and 90 per cent of those who are phobic of animals, insects and the natural environment, including fear of storms, the dark or water. Fear of a specific situation, such as public transport, tunnels, bridges, flying or enclosed spaces, is similarly skewed towards women. Blood-injection-injury phobias and fear of heights are more equally distributed, but still between 55 and 70 per cent of phobics are women.

Social phobias buck this trend. More women than men in the general population report having social anxiety but in the clinic the ratio of the sexes is approximately equal and, if anything, includes more men.

Put together, these ratios provide few clear clues to possible reasons behind the sex differences. Both agoraphobia and specific phobias are more common in women; social phobias are possibly more common in men.

The ratios challenge some of the most well-versed explanations for the differences between the sexes. An enduring idea is that women may be more willing to seek medical help for mental disorders, and more likely to be given a diagnosis by the physician they see. So the sex ratio in the clinic could be quite different from that in the community. So if there are equal numbers of male and female agoraphobics in the population, statistically you might expect that three times as many women as men show up and are diagnosed in the clinic. This seems unlikely. Firstly, far more women than men in the community have symptoms of agoraphobia. Secondly, the ratios suggest that men are perfectly willing to seek help for social phobia, so it seems unlikely they would not admit to agoraphobic symptoms.

Another approach is to blame society, which has traditionally valued different traits in men and women, boys and girls.
Fear of animals, insects and situations may be tolerated or
even covertly encouraged in a girl but are more likely to be
challenged and teased out of a boy. A woman who stays at
home keeping house for her husband is less likely to excite
comment than a man who does the same for his wife. Men
have traditionally felt more pressure to perform than women
in everything from sex and dating to work and home life,
and are more likely to lose face with their peers, family or
friends if they baulk at situations through fear. If over a
lifetime they are pushed into situations they would rather
avoid, they may be having years of low-level, real-life, on-
going behaviour therapy, and it has been used to explain why
men have fewer phobias than women. It does not, however,
explain why men have more social phobia than women.

The confusing sex ratios described in social phobia are
difficult to explain. It could be that women are more ready
than men to admit to the milder social anxieties seen in the
community, while severe social phobia is more disabling for
men than women so that more actually come forward for
help. However, this rather contradicts the explanations given
for agoraphobia.

The influence of female hormones runs into similar problems. If oestrogen and progesterone are involved in anxiety,
they are blunt instruments and are likely to influence general
anxiety levels. Times of hormonal turmoil might then be
expected to have most impact on the complex phobias of
agoraphobia and social phobia which are closely related to
general anxiety. It does not, therefore, explain why agoraphobia is so much more common in women while social
phobia appears less frequent than in men. And it is difficult

to see why these hormones would have anything to do with fears of insects and animals which tend to start in young children, long before surging hormones take effect.

As with so many putative causes of phobias, none of these ideas will provide the whole story and all could contain some truth. It is surprising, though, that the gender difference in phobias has been largely ignored for so long. Agoraphobia is a debilitating condition which hits people regardless of their social and economic status, job or profession. Previous chapters are testimony to the difficulties inherent in trying to pinpoint similarities in the lives or biology of people with agoraphobia. Yet, very obviously, almost three out of four sufferers are women. That makes female sex a more powerful risk factor than, for example, the genes for panic disorder which scientists are so carefully searching for. Gynaecology could in the end provide more answers than psychology.

The helpful effect of pregnancy reported at the beginning of this chapter prompts a spate of explanations. Pregnancy brings with it many biological, physical and social changes, some of which might be expected to make things worse, according to theories earlier in this book. Metabolic rate and breathing speed up as a normal part of pregnancy. If, as discussed in chapter 6, people who panic are especially sensitive to bodily sensations, these changes could even precipitate panic. On the other hand, pregnancy also provides a wholehearted distraction from normal concerns. Nicola, whose social phobia was described in chapter 3, thought one reason she improved so much while expecting her daughter was that she had an alternative explanation for any odd behaviour. Nobody would think her weird if she wanted to leave a meeting abruptly; she could blame her pregnancy, not her phobia. More generally, pregnancy can give an enormous boost to a woman's self-esteem and sense of purpose, which could improve psychological well-being and decrease residual anxiety.

George and his colleagues suggest that pregnancy hormones might have a more specific, biological effect on panic disorder. Chapter 4 discussed how easily people with panic can be aroused. Some have an exaggerated response to such basic physical challenges as standing up from a chair. Their blood pressure and heart rate increase far more than expected. Pregnancy hormones may blunt the reaction to these challenges and prevent the surge in blood pressure and heart rate.

Another possibility is that chemicals created when the hormone progesterone is broken down in the body may have a specific anti-anxiety effect. Other researchers have found similarities between these natural chemicals and barbiturates, a class of drugs used to treat anxiety. It may be that progesterone is broken down into a natural and effective anti-anxiety chemical.

Two female hormones, oestrogen and progesterone, have crucial roles to play during and after pregnancy and their levels change at different stages. Progesterone is dominant throughout most of pregnancy but levels of oestrogen start to rise towards the end. The stories at the beginning of this chapter hint at the involvement of the female hormones, particularly progesterone. But progesterone and oestrogen are also involved at other biological landmarks in women's lives, and if they are behind the impact of pregnancy on panic, logically, they should also have an impact around the time of the menopause.

In fact, the *American Journal of Psychiatry* has published

a series of letters describing the experiences of individual women. In 1992, one letter suggested a link between hysterectomy and panic disorder. Two years later, other researchers in the US, led by Mark Dembert in Bethesda, Maryland, pointed out that women who have this operation are usually given hormone replacement therapy (HRT). They suggested it could be the female sex hormone, oestrogen, in the HRT, rather than the operation, which is responsible for the panic disorder.

One of their patients, Ms A, had her womb and ovaries removed when she was thirty-seven and started taking oestrogen shortly afterwards. A month later, she was having frequent, full-scale panic attacks for the first time. She was found to have a disorder of the thyroid gland called Graves' disease, which can give people panic-like symptoms. However, after this was successfully treated, the panic attacks continued. They only stopped when she gave up the oestrogen. A few months later, her gynaecologist prescribed oestrogen again, this time at a lower dose. Once again, she started having panic attacks within days. She put up with this for a short time before stopping the oestrogen. Her panic attacks subsided within ten days and one year later she remained happy and confident. For this woman, at least, replacement oestrogen seemed to be a direct cause of panic.

Hormone replacement therapy is widely used by women around the time of the menopause but psychiatrists are unlikely to recognise it as a possible cause of panic disorder. They will reach instinctively for anti-panic drugs or psychotherapy and it may never cross their minds to consider stopping the HRT.

Dembert and his colleagues were backed by another letter a year later. A Dutch group led by Eric Griez wrote that pregnancy is commonly thought to protect against panic attacks because levels of the opposing female hormone, progesterone, rise. Like George, they witnessed the calming effects of pregnancy. A woman who had had panic attacks and agoraphobia for years postponed treatment because she was pregnant. Nine weeks into the pregnancy, her panic stopped. It started again two weeks before delivery. The same thing happened in her second pregnancy: her panic stopped in the third month, but started again four weeks before she gave birth.

Griez believes that the natural progesterones in pregnancy are responsible for reducing panic. Most panic in pregnancy happens in the last few months and they suggest this is caused by the change in the balance of hormones. Levels of oestrogen start to rise more quickly than progesterone after the fifth month and the late surge of oestrogen may be responsible for the return of panic. He was encouraged by Mark Dembert's observation that oestrogen replacement therapy may also make menopausal women susceptible to panic.

## The Numbers Game

So far, so good. However, while a handful of reports about individuals' experiences makes a good starting point for research, it is not firm evidence. Scientific literature is vulnerable to some of the same biases as newspaper reports. Everyone loves a good story and it is a human trait to be interested in the dramatic or the unusual. This is not meant to suggest that any of the case reports is untrue, only that they may be the exception rather than the rule. If a woman's panic disorder remains unchanged throughout pregnancy, her psychiatrist is unlikely to be motivated to write up this mundane event and send it off to a journal. Equally, staff at the journal's office are unlikely to be interested enough to publish the story.

Studies that include tens or hundreds of people are essential before firm conclusions are drawn and this work is in its infancy. Jerrold Rosenbaum, whose work on personality was discussed in the last chapter, set out to look at groups of women going through pregnancy and the menopause. The results make confusing reading.

Rosenbaum and his colleagues studied forty-nine women who had panic disorder for years before becoming pregnant. Most of the women were referred to psychiatric units because of worries that the foetus could be harmed by untreated panic symptoms or by anti-panic drugs. The women had severe panic disorder, but many were highly motivated to come off the drugs while pregnant.

In contrast to the women described at the beginning of the chapter, this group did not sail through their pregnancies. Three out of five had no change at all in their panic symptoms. One in five got better and one in five got worse. Pregnancy had no clear-cut effect on panic. But closer examination of the different groups of women showed that those with the mildest panic disorder were the ones most likely to improve. Women who had particularly severe panic disorder before they got pregnant were most likely to get worse.

Most of the women taking drugs attempted to come off them before conceiving and, as expected, the less severe the panic disorder, the more likely the women were to succeed. A few were unable to stop before they conceived but were successful once they knew they were pregnant. Having a confirmed pregnancy may have increased their incentive or they may have been helped by their hormones' calming influence.

Rosenbaum is not saying that pregnancy has no effect on panic, just that the effect is unpredictable. He wants to be able to predict in advance which women are most likely to get better, and which worse, so that the women and their doctors are able to work out the best approach to take during pregnancy.

His group also studied women after delivery, since the post-partum period has traditionally been viewed as a high-risk time for the development of psychiatric disorders. They followed forty women who had panic disorder before they conceived and found that, as in pregnancy, most of the women – more than three in five – showed no change at all in their psychiatric well-being. Most of the others got worse: only three out of forty showed any signs of improvement.

This study tells us little about the undiluted effects of the aftermath of pregnancy because most of the women who stayed the same or improved were taking anti-panic drugs. Most of those who got worse were not taking drugs. The conclusion the researchers drew was that even mild symptoms of anxiety during pregnancy should serve as a warning that women may need to start drug treatment soon after their child is born in order to head off the development of severe symptoms.

These studies were criticised by Donald Klein, whose suffocation false alarm theory was covered in chapter 4. His own work was more in accord with folklore. He found pregnancy more beneficial, and the post-partum period worse, than Rosenbaum's group. Women who breastfed their babies were also less likely to get worse postnatally. Klein suggested that pregnancy hormones could act via the respiratory system.

Progesterone levels during pregnancy reduce the stimulus to breathe; after birth, this stimulus is increased.

Klein complained that some of the women in Rosenbaum's group may have been seen by prescription-happy doctors, since they appeared to have started taking drugs without having deteriorated. This, he conceded, could have been because the study interviewed women about their medication after changes had been made. In response, Rosenbaum's group embarked on another study, this time interviewing women in advance.

A preliminary study was published two years later and included only ten women but found similar results. Rosenbaum's women did not bloom during pregnancy. Seven of the ten met criteria for panic disorder at the three visits during pregnancy, and another two met the criteria at some visits but not others. Only one was panic-free throughout pregnancy. This case, of course, is the one other doctors might have chosen to write up as a case report. But this one woman's happy experience sounds less impressive than it might when it is shown against the backdrop of another nine women getting little, if any, relief from panic during pregnancy.

Four of the ten women continued on their normal anti-panic drugs, or even increased them, and three of these women continued to have panic disorder. The researchers point out that if pregnancy were protective, you would not expect so many women to take so many drugs, particularly when most were concerned about effects the drugs could have on the foetus. Three other women tried to stop taking their treatment but all continued to have panic disorder. Pregnancy did not appear to make it any easier for women to stop taking drugs.

The story was much the same after the women had given

birth. Most increased their treatment in the three months following delivery but nine out of ten continued to have panic disorder despite the treatment. This study is a step forward from case reports but as it only included ten women much larger-scale work is needed to get firmer answers

Confusingly, exactly the opposite finding came from a study in South Carolina. Researchers wrote to women who had become pregnant after being diagnosed with panic disorder. Most who replied, fourteen out of twenty-two, said their symptoms had improved during pregnancy. However, the women had varied experiences. Some felt better only at certain stages of pregnancy, others did not feel better at all.

The apparent contradiction between this study and Rosenbaum's work could be because the two groups of women studied were quite different. The women in South Carolina could have had a milder form of panic disorder than those in Boston, who were all being seen at a hospital psychiatric clinic. It is well known that women who have had spells of depression in the past are at increased risk of becoming depressed after giving birth. Similarly, Rosenbaum's study suggests that more severe panic disorder before conception can increase a woman's chances of continuing to have symptoms during pregnancy. These women might be a particularly vulnerable group.

This could mean that anti-panic drugs are the best solution for women with the most ferocious panic disorder. The risk of untreated panic may be greater than risks associated with exposing the foetus to drugs. The same may not be true for women with less severe panic disorder, who may be able to get through pregnancy without taking drugs, perhaps helped by cognitive and behavioural therapy.

Rosenbaum believes that fluctuating hormone levels

influence anxiety levels but says it is still unclear which hormones are involved and what their effects are. Women take artificial hormones at different times of their lives and for a variety of reasons, including control of menstruation, contraception, ease of menopausal symptoms or postmenopausally to reduce the risk of conditions such as osteoporosis. The effects of the hormones are difficult to predict. Some women get better and some get worse on artificial oestrogen. Some get spectacularly worse with synthetic progesterone.

The times when women take hormones may be emotionally loaded for other reasons: pregnancy and the menopause are highly significant events in women's lives. To Rosenbaum, however, this is not a sufficient explanation. If a woman has had mild menopausal symptoms for months before starting on hormone therapy, and is then quite suddenly unable to bear her anxiety, it is not unreasonable to think the synthetic progesterone has triggered the problem. Natural progesterone seems to have a beneficial effect for some women in pregnancy as reported earlier. It also sometimes improves anxiety associated with the menstrual cycle. It is strange, then, that synthetic progesterone can trigger this dramatic downturn. But synthetic progesterone could have a different biology from the natural stuff. If the body breaks it down into compounds with a slightly different shape, these molecules may not be able to fool the body's biology that they are natural barbiturates.

Taken together, the work so far suggests that the female hormones oestrogen and progesterone can have a profound impact on anxiety levels. We cannot predict what the effect of each of the hormones will be, nor make more than an educated guess as to how the hormones act. Neither do we have much idea what to do about it.

## Teamwork

The treatment received by a woman with anxiety related to her fluctuating hormone levels is likely to be heavily influenced by her doctor's discipline. A woman who has panic disorder before she becomes pregnant may be used to seeing a psychiatrist and continue to do so while pregnant. She is likely to receive anti-panic medication. A woman who becomes severely anxious after taking synthetic progesterone for menopausal symptoms may never have seen a psychiatrist in her life and would be more likely to go back to the gynaecologist. She may receive a different hormonal treatment.

For there to be real progress, gynaecologists need to team up with psychiatrists to learn about these conditions. London gynaecologist John Studd recently called for gynaecologists and psychiatrists to collaborate in clinical trials for the treatment of female depression. He insists that premenstrual, postnatal and premenopausal depression is best treated with oestrogen. Psychiatric and feminist 'dogma' has blamed psychological and environmental factors for female depression, he says, leading to the overuse of anti-psychotic drugs.

In the US, some such collaborative work has been done, though with the anti-depressant drug, sertraline. This drug increases the brain's supply of the calming chemical messenger serotonin and reduces depression and anxiety. A study included 200 women with severe premenstrual syndrome who were debilitated every month with anxiety, depression, mood swings, impaired productivity at work, relationship difficulties and reduced social activity. They took sertraline for two or three months and more than half improved significantly, though one in three did not. The researchers concluded that family doctors could offer this treatment to women with severe premenstrual syndrome. It demonstrates that even if the cause of a problem almost certainly comes down to hormones, the solution may not have to.

Projects that combine the expertise of gynaecologists and psychiatrists are likely to be productive in future but are only just beginning. With emerging evidence linking anxiety, panic and agoraphobia to the menstrual cycle, pregnancy and the menopause, these events cannot remain under the sole domain of gynaecologists.

Psychologists and psychiatrists have shown sporadic interest in female hormones over the years. In 1975, an interesting project was carried out by London psychologist, Doreen Asso. She speculated that women might be more or less likely to aquire specific phobias at different times of the month. She gathered two groups of women, some phobic, some not, and tested their reactions to the mildly unpleasant noise of a car's horn sounded shortly after they saw a blue light. Like Pavlov's dogs, after several repeats of this experiment women develop a so-called conditioned response and start to react to the blue light alone.

Women who were not phobic were significantly more likely to develop this conditioned response if they were premenstrual compared to being mid cycle. Women who were phobic were more likely to develop the response at any time.

Asso concluded that non-phobic women may have a short phase each month when they are particularly vulnerable to developing undesirable responses. Women prone to phobias are even more susceptible at this time. She suggests that women are excessively aroused in the four days before a period when levels of oestrogen and progesterone both decline. The general arousal makes it more likely that an unpleasant event will make a lasting impression. This window of vulnerability every month could even be, she suggests, one of the reasons why women have more phobic symptoms than men.

Women interviewed for this book would support Asso's theory. Wendy, whose fear of winds was described earlier, believes her phobia took hold partly because on the day the roof blew off, she was premenstrual. 'I was at a low ebb with PMT, I was a bit run down. It just clicked,' she said.

Another woman, Adele, whose agoraphobia was success-fully treated by behaviour therapy, had her first panic attack just before a period. She had noticed previously that she often felt dizzy a couple of days before she was due. 'It was funny. If I was running for the bus, I would feel as if I was still running, even after I had stopped. And I felt as if some-one had put itching powder in my coat; my arms felt prickly.'

The panic attack happened while she was on holiday. 'I kept saying to my boyfriend "Keep walking me up and down the side streets." I felt as if, if I sat down, I would die. He had to keep walking to keep me alive.' Adele developed agoraphobia and was housebound for three and a half years. Now recovered, she often still feels dizzy before a period, but will not allow herself to be overtaken by it.

Others, unprompted, mentioned their hormones. Wendy got worse during the menopause and improved afterwards, as did Elizabeth. Nicola (and Mary – described in the next chapter) improved during pregnancy, Elizabeth got worse. However flimsy the scientific evidence for the impact of hor-mones on panic, women themselves are in little doubt about its relevence.

## Migraine and Marriage

Another approach is to look at other medical conditions that are more common in women than men and see whether anything can be learnt from the comparison. For example, an opportunistic study in the Coral Sea followed a tropical cyclone, code-named Justin, off north-eastern Australia in March 1997. An ocean liner carrying 1,350 passengers and crew was caught in rough seas and gale-force winds. Three enterprising family doctors decided to monitor seasickness on board.

They found that passengers were much more vulnerable than crew members, which is not surprising, since people who suffer badly from seasickness are unlikely to choose a career on a ship. But women were also three times more likely to be badly seasick than men. The Australian doctors linked this sex difference with migraine, also more common in women, and suggested there could be a common cause. The final chapters in this book look at possible links between motion sickness and migraine with phobias. The study on the Coral Sea is a one-off, but it also brings in gender as a risk factor for all three conditions.

Physical and hormonal differences between the sexes may make a fruitful area of study in future but the sexes' difference in psychological outlook should not be ignored. It would be as wrong to dismiss this body of evidence as it has been to ignore hormonal influences for so long.

One persistent theme in the research literature is that even if men and women have the same symptoms, they may react differently to them. This is true throughout psychiatry. Bob Cloninger, whose work on personality was discussed in the last chapter, found that the same underlying psychological disturbance could look quite different in men and women. Where men turn to crime and alcohol abuse, women may develop physical illnesses, especially pain in the head, joints or abdomen.

Attitudes to alcohol have been quite different across the sex divide. People with agoraphobia are more likely than most to turn to alcohol and one in five are alcoholic. This is almost twice as many as in the general population (according to the American ECA study, mentioned in the introduction). But it is not helpful to lump the sexes together in these comparisons because there are important differences. Agoraphobia is more common in women and alcoholism is more common in men. Canadian researchers looked at a mixed group of people with panic disorder and agoraphobia. They found that the men drank more alcohol and were more likely to believe that it helped them cope with their anxiety. The more severe the men's anxiety, particularly social anxiety, the more they drank.

Women were much less likely to react in this way in the Canadian study, but alcohol is still firmly linked with agoraphobia. A study of female alcoholics found that almost one in three had panic disorder with agoraphobia. Alcoholism is relatively rare among women and the researchers concluded that women with agoraphobia might represent a high percentage of female alcoholics.

Some of this difference is likely to be determined by culture. Heavy drinking has traditionally been seen as more acceptable in men than women. Men can drink far more than women before exciting comment, let alone a diagnosis of alcoholism. Heavy drinking habits could be well-established in men before anyone notices they are out of the ordinary.

Cultural norms are intricately bound in with agoraphobia in other ways. The *DSM* urges doctors to remember that some cultures restrict women's freedom to take part in public life. A woman who never leaves home, bowing to the dictates of her culture, does not have agoraphobia.

On a much smaller scale, different individual expectations still guide the lives of men and women. Women are far more likely to stay at home looking after house and children, and this is not an easy option. Women with at least one child at primary school are more prone to anxiety and depression if they stay at home, rather than go out to work, according to a study at Heriot-Watt University.

Anxiety has a complicated effect on relationships. In a surprising number of couples, both have an anxiety disorder. The partners may have chosen each other because they identify with each other's neuroses, or perhaps illness in one triggers anxiety in the other. This is important in treatment. If one person's anxiety has made the other ill, then successful treatment should improve both partners and the relationship. If, however, anxiety is a core feature in the attraction between partners, treatment of one might be strongly resisted by the other, and if successful, could undermine or even destroy a relationship.

London psychiatrist Julian Hafner studied the husbands of women with agoraphobia in an attempt to answer this question. He found that two-thirds of the men would not be diagnosed as having an anxiety problem and appeared generally satisfied with their marriages. Any symptoms the men had were not easily linked with the severity or duration of their wife's agoraphobia. The impetus to get treatment came primarily from the women, who were much more dissatisfied with themselves than their husbands were. Hafner concluded that the anxiety might indeed be an important part of many of the marriages. The men were happier than the women with the status quo and might be expected to resist treatment. An A-frame relationship is one in which each partner props the other up, like the sides of the letter A. The danger is that if one partner learns to stand up on their own, the other loses their role within the partnership and the marriage falters as a result.

In order to avoid this, Hafner said that men could be invited to take part in their wives' treatment in the hope of helping them make any changes that might be necessary. This could help the woman get through treatment and it might, presumably, also help keep the marriage together.

While other studies have agreed with Hafner's supposition that successful treatment for agoraphobia can undermine a marriage, Isaac Marks thinks it is nonsense. He points out that most agoraphobics are treated when they are at the prime age for separation and divorce. Given that about one in three marriages end in divorce in the general population, he says, it is hardly surprising that some marriages go wrong during treatment.

Marks's own work found that reduction in phobias either improved marriages or left them as they were; certainly they did not get worse. And the better the marriage, sex, social life and work adjustment before treatment, the more likely the couple were to benefit as a result. This was true regardless of the severity of the agoraphobia and Marks believes it may have more to do with the couple's ability to adapt to change. If they have already demonstrated that they can solve other problems, they are more likely to be able to cope with the upheaval that could result from successful treatment for agoraphobia.

Despite Marks's and Hafner's different theoretical stand-points, the practical outcome may be similar. Marks has long advocated the involvement of a spouse as co-therapist in behaviour therapy, as a way of helping their partner to carry out the tasks usually set as homework between sessions. David Barlow, in Albany, New York State, has shown that including the partner in treatment for agoraphobia makes it decidedly more effective. This could be because the spouse helps with the homework, but might also be that treating a couple rather than an individual reduces marital discord.

Barlow examined changes within marriages throughout the course of treatment, looking particularly at marital happiness and communication. Marital happiness did not seem to influence treatment very much, surprisingly, but communication did. Those who did well in treatment rated their partners as more communicative than those who did not respond to treatment. The finding was strong enough for Barlow to recommend specific communication treatment for some couples before they embark on exposure therapy.

Home life, often so different for men and women, can have a profound impact on the success of treatment. It may also influence the development of agoraphobia in the first place, but not always in the expected direction. Myths abound on the theme that middle age is a difficult time for women. The 'empty nest syndrome', whereby women lose their nur-turing role when children leave home, has been assumed to knock their self-esteem. However, a London researcher followed more than 100 women over seven years, from their thirties through to their mid forties, and found quite the opposite. Women with high self-esteem to start with retained it through the duration of the study. Those with low self-esteem at the beginning also improved markedly as the years went on.

The researcher, Bernice Andrews, found that the positive changes in self-esteem could be attributed to positive changes in the women's lives. Important events included promotion at work, leaving an unhappy marriage, forming new and more rewarding attachments and getting on better with their children. Another, larger study found similar results but came to a different conclusion. London researchers carried out psychiatric assessments of 9,700 women aged over fifty-five. They had much lower levels of depression than younger women – lower, in fact, than men. This study concluded that social explanations for the reduction in depression were inadequate and suggested the drop in women's problems might be directly related to the passage through the menopause.

## Girl Power

Carol Nadelson, from the New England Medical Center in Boston, specialises in the impact of gender on many aspects of psychiatry and psychology. She believes that cultural and social expectations are likely to be highly relevant in a condition like agoraphobia. Biology will also play a role, she says, but without looking at the difference in the lives of men and women, it is impossible to fully understand why so many more women than men have agoraphobia.

Her starting point is to challenge directly some of the assumptions underlying a diagnosis of agoraphobia. Agoraphobia is an irrational fear of leaving home, walking through crowds, using public transport. But there is a fine line between what is rational and what is not. Life might actually be more dangerous for women. They are more vulnerable to physical and sexual abuse than men. Women are the ones who get pregnant, and an unwanted pregnancy is a more serious issue for the woman involved than the man. They also tend to look after small children; anyone with this responsibility might be especially aware of dangers on the streets and more loath than usual to go out. Is that agoraphobia, or is it due caution?

Women, particularly single mothers, are also more likely than men to be financially worse off. If they live in poorer, rougher neighbourhoods, where streets are inadequately lit at night, how wise would they be to go out?

Furthermore, traits inextricably linked with agoraphobia might be endemic among women with *and* without anxiety disorders. Lack of assertiveness and aggression, and a reluctance to take risks are commonly seen even among highly successful career women. Competence and intelligence are still seen as a liability in the dating game, says Nadelson, and women tend to value relationships more highly than anything else in their lives. She believes that the self-doubt which plagues so many women, and the sense that it is improper to be too assertive, has something to do with a fear that they will be rejected. Losing a relationship can feel to many women like a more serious risk than almost anything else.

This conditioning starts at a young age. Girls tend to grow up protected by families where boys are encouraged to express physical aggression and become more comfortable with it. Girls' tendency to focus on relationships is often reinforced by their upbringing, being encouraged to stay close to home. A boy who gets a cut lip in a game is heroic, and any scar will make him more masculine and attractive. If a girl has the same injury, everyone worries that the scar will mar her life.

Ambitious behaviour, applauded in men, may still not be seen as acceptable in women. Women abound at the entry level of careers but fall away higher up the ladder. In no profession have women gained the power and recognition that would have been expected if you looked at the calibre of people entering it twenty-five years ago. In medicine, the number of women who are heads of department or deans of medical schools is as low as it has ever been. The occasional woman has always made it to the top, but women still do not succeed in great numbers. Nadelson says there is always a reason and an excuse on both sides. Women, faced with an overwhelmingly male hierarchy, often back off, unwilling to take more than their share of rejection and disappointment.

Finally, this is changing. Girls and young women have a greater sense of autonomy and assertiveness than ever before. They are happy taking on physical, lifestyle and career choices unthinkable even twenty years ago. They are more active and confident than their mothers could ever be. Expectations are changing, if slowly. Fewer doors are closed to women and society is starting to allow them to be successful, intelligent, ambitious *and* attractive.

This could have a tremendous impact on phobias. Women have different hormones from men and are always going to be the ones having babies, periods and the menopause. Some of the differences between the sexes are never going to change. But as society shifts its expectations and allows or even demands the same behaviour from girls as from boys, girls will cease to express some of the traits that provide such a fertile breeding ground for phobias. Girls who have grown up without restrictions or cosseting may have a sharper will to achieve in all aspects of their lives. As the status of men and women becomes more equal, women may become more phobia resistant. Girls who grow up wanting and expecting to go out and achieve, will in future be far less likely to suffer from the fears and phobias so commonplace among women today.

## CHAPTER 9

# Light and Electromagnetism

## Mary's Story

A woman developed a severe, uncharacteristic and extremely troublesome phobia of water. It started after she was bitten by a dog, which subsequently died, in India. She went to see her doctor and was told she had an anxiety disorder, and a specific fear of water – hydrophobia. The doctor made suggestions but she did not improve; hardly surprising, because she had all rabies' classic symptoms and was not receiving any treatment for it.

This story was told on a radio programme a few years ago, and it struck a chord with at least one listener. Mary Dwarka was housebound with agoraphobia at the time. She, too, had been told that her condition was anxiety-based, in her case linked with panic disorder. She had seen a psychiatrist and been diagnosed as mentally ill.

Dwarka did not recognise herself in the descriptions she was receiving from the health professionals. Her depression and panic seemed to be the result – not the cause – of being stuck at home. She felt her problem was more strongly linked with travel than with anxiety and started to wonder if she, like the woman on the radio, had been given a knee-jerk diagnosis when in fact the problem's root lay elsewhere.

With two children to look after on her own, it was an extremely hard time for Dwarka, but she was determined to make sense of what was happening to her. She had worked for years as a civil servant and had no medical or scientific knowledge, but believed that her observations about her own condition would reflect the experience of many others with agoraphobia. She scoured the media for items on fear, phobias and any related subject, however tentatively linked. She wrote to the scientists interviewed and received some helpful replies. She carried out a detailed survey of 175 women with agoraphobia attending the self-help groups. And, over a number of years she developed an original and all-encompassing theory about agoraphobia which takes into account some of the most exciting and controversial fields of research today. In turn, researchers have praised her work and some have gone on to explore her suggestions.

Dwarka believes that agoraphobia is, at heart, a travel disorder. She rejects the literal translation of agoraphobia as too limiting. It is, rather, a problem of getting from A to B. Travelling often takes us through open spaces, of course, but she believes the fear is rooted in other aspects, notably our ability to navigate.

She agrees with the evolutionists' arguments earlier in the book, but she takes them a step further in relation to agoraphobia. The crunch point, she says, is that when we set out from home, we want to know that we will be able to find our way back again. For our Stone Age ancestors, who could not survive alone in the wilds, this would have been a matter of life or death. They roamed for great distances, hunting and searching for food, but they always knew where the home base lay. It was essential for them and for their offspring that they made it back.

However they found their way – they could have used familiar landmarks, the sun, moon and stars, light and the earth's geomagnetic field – those that were efficient navigators had a survival advantage. Anyone unable to stray too far from base would have missed out on the best pickings and their children could have starved; anyone who got lost probably perished. Those who could navigate were most likely to survive and pass this advantage on to their offspring.

Our environment and lifestyle have since changed beyond recognition and herein lies the problem. If we are perfectly suited to travelling by foot, and navigating according to natural cues, what use is it as we commute fifty miles each way by high-speed train? Or when we are hurtling down a motorway at top speed? Or stuck on a packed, stalled underground train, going nowhere?

Dwarka believes the problem for agoraphobics is not that they actually get lost, but that their internal, natural sense of navigation is so disrupted by modern transport systems that they feel horribly disorientated. It could be due to the magnetic fields set up by underground or electric train systems, the artificial lighting on almost all public transport or even just the sheer speed and distances covered. Whatever the underlying reason, it is the disorientation, rather than anxiety, that triggers the panic reaction.

Agoraphobia is too often lumped together with panic disorder. The American ECA study found that more than three times as many people have agoraphobia as have panic disorder. She says they are not the same thing, and the sooner we start considering agoraphobia as something other than an extension of panic, the sooner we are likely to make progress.

Disorientation could be made worse by the way travelling confines us. Journeys by train or car can lack an escape route: once the train has left the station, you cannot get off; once you have passed a motorway junction, you have to continue along the road. Fear of such journeys is classed as agoraphobia, but the problem may be the lack of an escape. Even walking in a park, can make you feel confined if you have to stick to the path. Agoraphobia and claustrophobia, the fear of enclosed spaces, appear to be opposites, but may in reality be the same thing. And they could both hark back to the Stone Age, when journeys through wild countryside would not have been contained in the same way as ours so often are.

Consideration of our ancestors may also explain another statistic: why women are two or three times more likely to have agoraphobia. Men usually hunted further from base than women, who were responsible for looking after children and could not easily travel as far. Perhaps men developed a surer navigational sense. Or maybe women have as good a natural sense, but find it disrupted by raging hormones from time to time.

Dwarka had mild agoraphobia for years before she became pregnant with her second child. When she became pregnant, like some of the women in the previous chapter, she had a total reprieve and could travel to work or to social events without the dread she had become accustomed to. It was months after the birth of her baby, when her agoraphobia came back, and then it was with such a vengeance that it kept her at home for years.

Her own survey found – not unlike Rosenbaum in chapter 8 – that two in five women similarly felt much better when pregnant. Most women – 73 per cent – also said symptoms were worse at different times of their monthly cycle, usually when premenstrual. Agoraphobia ebbs and flows.

This is curious. By contrast someone afraid of snakes is pretty much always afraid, in any environment. A snake is frightening whether in the flesh, in a film or in a picture. This would have made sense thousands of years ago. Snakes are constantly dangerous and it pays to avoid them at all times. But people with agoraphobia know that going out can bring pleasure as well as pain. Beautiful outdoor scenes on television provoke no reaction. Pictures of the American prairies are not upsetting. Many agoraphobics have good days, when they can travel much as normal, and bad days when they are housebound. So – what determines when they are afraid?

Dwarka believes that the reasons are internal and, again, rooted in ancient times. In a constant external environment, women's vulnerability none the less fluctuated. At certain times of the month they may have been less able to navigate. Pregnancy may alter the hormonal balance for years, and this may have been adaptive if small children were best cared for by a mother disinclined to stray far from home. Hormones that curtailed women's movements at critical times may have kept them, and their offspring, alive.

In a nutshell, Dwarka is suggesting that agoraphobia is a travel disorder, not an anxiety problem. It stems from the gulf between Stone Age voyages on foot, and modern transport. Our ancestors may have relied on light and the earth's magnetic field to navigate; our vestigial navigational sense is thwarted by trains and motorways. This can disorientate anyone at any time, but women may be especially vulnerable because of hormones that surge cyclically and after pregnancy.

On first reading, some of this sounds implausible. But Mary Dwarka is not a lone voice. Researchers in mainstream medicine have devoted substantial resources to some of these suggestions. Just how strong is the evidence?

## Magnets and Migration

It must have been an odd sight. The van was driven some distance along winding roads and when the driver finally stopped, it was apparent that the passengers were blindfolded. They were asked to point in the direction of home, and their responses were noted down. Then it was off with the blindfolds and time to go. End of experiment.

Another group looked even stranger, with bar magnets strapped to the sides of their heads. One by one they were led to a designated place in a strange wood and told to make their own way back. Their progress was carefully monitored.

These unlikely sounding experiments give some of the most direct proof there is for Dwarka's hypothesis. They were conducted in the 1980s and early 1990s by Robin Baker, then a zoologist at Manchester University, and they examined the way we navigate.

The blindfolded volunteers in the van could not make a mental note of landmarks, or work out their orientation from the position of the sun. They could see nothing, but were still better than could be expected by chance at determining the direction they had come from. They were getting information from somewhere. The bar magnet experiment in the wood suggested that this information could be electromagnetic. The magnets interfered with a small patch of the earth's geomagnetic field and volunteers wearing them had more problems finding their way than those wearing a dummy bar.

The impact of the magnet depended on the weather. When the sun was low in the sky and strongly directional, there was little difference in performance whether or not the magnet was real. But when the sky was overcast or the sun high up in the sky, giving few directional hints, the volunteers wearing active magnets had most problems orientating themselves. It seemed the sun was the navigational instrument of choice and geomagnetic information a last resort when the sun could not be used. Wearing a vertical bar magnet in front of the ear for about ten minutes disrupted performance for an hour but total recovery could take two days. Different magnets improved volunteers' performances.

Baker and other groups have conducted many variations on the theme of these experiments. They have used electromagnetic helmets or coils, they have tried treating their volunteers beforehand in a magnetic field. Sometimes they asked volunteers to point in the direction of home, at other times to draw an arrow on a piece of paper. In so-called chair experiments, volunteers wore a blindfold and earmuffs and sat on a swivel chair. They were spun in one direction, then the other and, every time the chair stopped, were asked to guess which compass direction they were facing.

As far as Baker is concerned, the findings point in one direction. Normal, healthy people rely on both light and electromagnetism to find their way around. We prefer to use light, but when that is not an option, can make use of non-visual information. Part of this, at least, comes from the magnetic field around us.

Given the existing work on birds and animals, this is perhaps not as extraordinary as it sounds. Migrating birds travel thousands of miles and routinely nest in exactly the same spot year after year. Homing pigeons, released hundreds of miles away, come back loyally to their owners. Seals migrate through the depths of the ocean and regularly return to the same mate and the same rock. We accept that fish, birds and mammals use magnetic fields to orientate themselves. Exactly how they detect these fields is not known, but scientists have made many suggestions. Fish might have electroreceptors; animals might use their inner ear or their eyes. Baker proposed that all creatures have an organ in the brain, a so-called magnetoreceptor, dedicated to picking up magnetic fields. It could be the link between magnetism outside the body and its effect inside, the site at which external magnetic fields are converted into nerve impulses.

Species as diverse as bees, homing pigeons and dolphins have all been shown to have magnetic particles of an iron oxide called magnetite in their brains. This could be the basis of Baker's magnetoreceptor and, with colleagues at Manchester, he set about looking for magnetite in humans. They took samples from different parts of the heads of people who had recently died, exposed them to a strong electromagnetic field and tested for magnetite and magnetic activity.

They found both in the sinuses, an area midway between the eyes but slightly lower and further back. It is technically called the sphenoid/ethnoid sinus complex, and lies close to the nasal bones and pineal gland. Significantly, it is the same area of the brain in which magnetite has been found in fish, mice, dolphins and birds. Baker's magnetoreceptor for humans and animals alike was not proven, but the idea was starting to stand up.

Elsewhere in science and medicine, electromagnetism receives patchy recognition. It has long been suspected of playing a role in the development of childhood cancers but numerous studies into clusters of leukaemia close to power lines or nuclear power stations have failed to prove the link. Radiation from computer screens has been blamed for miscarriages and birth defects in babies born to women working at terminals. Again, no link has been confirmed, but suspicions remain and standard regulations on computer screens have been tightened considerably over the years. The recent furore over the potential dangers of mobile phones emphasises our fear of radiation.

Military scientists have been interested in radiation for years but from a rather different point of view. In the 1970s, the US Navy dreamt up a project called Seafarer (Surface ELF Antenna For Addressing Remotely-deployed Receivers). The Navy wanted to bury an enormous radio antenna system a few feet below ground in order to diffuse extremely-low-frequency (ELF) radio waves through the biosphere. These radio waves travel through sea water and would allow the Navy to communicate with submarines deep down in the ocean. At the time, submarines could not communicate with base without coming up near the surface and putting up antennae. This was obviously more risky than being able to stay down in the depths of the ocean.

But early small-scale experiments by the Navy found that ELF had harmful effects. A group of people exposed to it had raised levels of a type of fat called triglycerides in their blood. Eleven men became significantly worse at simple arithmetic. An internal Navy report admitted that young ring-billed gulls might be disorientated by energised antennae. Others found that pigeons were deflected towards the antennae. Public opposition to the project grew on health and environmental grounds.

Other military projects have focused on radiation as a potential non-lethal weapon. Details are hard to come by, but informed observers have suggested that very low frequency noise is being examined to see whether it could be used against enemy soldiers. This could be especially useful for forces trying to keep the peace. A strong blast of the right frequency could leave rioters disorientated, perhaps, or even clutching their stomachs, vomiting and defecating uncontrollably. The best part of this is that, ideally, there are no side effects. Those targeted would suffer no subsequent ill effects.

Together, this work suggests that artificial electromagnetic fields in the environment could have an impact on our general health. It is a step towards backing Dwarka's ideas, but is not sufficiently linked to agoraphobia to be conclusive.

Baker, unfortunately, was unable to recruit agoraphobics into his studies; the nature of the experiments must have been horrifying for anyone nervous of venturing away from home. He studied people with dyslexia because they have spatial problems that may – according to Dwarka and others discussed in the next chapter – be compared with agoraphobia. Muddling letters and numbers, and confusing right and left, could be related to orientation. He found that people with dyslexia were almost indistinguishable from the general population but, interestingly, that their parents performed poorly in Baker's experiments. It raised the possibility that dyslexics could have parents who have trouble orientating themselves.

There is a shred of evidence, but no more, that women's hormones interfere with navigation. Female volunteers were significantly less accurate in Baker's chair experiments when premenstrual. However, many women report poor concentration and increased anxiety when premenstrual. The experiment could not prove that their dip in performance was due to an impaired magnetic sense. An impaired sense of humour might have had the same effect.

However, recent work using high-tech imaging techniques has demonstrated that women and men use different parts of their brain for navigation. German researchers examined the brain activity of men and women searching for a way out of a virtual-reality maze. They found that men primarily use the hippocampus, a very old part of the brain conserved through evolution. This suggests they use ancient instinctive mechanisms for navigation including, perhaps, electromagnetic fields. By contrast, women activated parts of the cortex, or thinking brain, which suggests they may rely more heavily on landmarks.

Sadly, Baker's research stopped before it got much further. When he shifted his interest from animals, birds and fish to humans, it became so contentious that he lost the general support of the scientific community. If evolution and genetics have taught us anything, it is not to assume that humans are different in fundamental ways from other living creatures. Baker, though, found that the idea of a magnetoreceptor in people was generally held to be unbelievable.

Other groups tried the same experiments and claimed not to get the same results. Baker became exasperated with the papers being published since, in his view, the new studies *were* confirming his own findings. 'The stark contrast between positive results and negative interpretations remains an unexplained feature of the literature on human magnetoreception,' he wrote in 1987. He eventually left his post at Manchester University and since then, the work has fallen by the wayside. Baker's concern is that many people are exposed to magnetic fields at work and we ought to know if this is likely to affect their health. Even at leisure, we could be at risk. Personal stereos set up a measurable electro-magnetic field around the head, he says, and young people in particular often wear them for hours on end. Further-more, if Dwarka's ideas about agoraphobia are right, it is a tragedy for such an original strand of thought to remain unexplored.

## Strip Lights and Seizures

It was difficult to hold a conversation with Maria. Every few minutes, her eyes would glaze over and her concentration would evaporate. Then she would be back, without regis-tering what had happened, and her companion would have to decide whether or not to repeat what they had just said.

In some ways, conversations were the least of Maria's prob-lems. She had pattern-sensitive epilepsy and whenever she saw stripes she would have an 'absence' and momentarily lose awareness. Stripes are everywhere, of course, and she was having up to twenty-two absences an hour. She could not cross a road unaided, and had to be monitored even while having a bath.

Research psychologist Arnold Wilkins was working with Maria while on sabbatical at the Montreal Neurological Insti-tute in Canada. He was studying the effects of fluorescent lights on several different medical conditions, including epi-lepsy. He devised a pair of spectacles with one frosted lens which reduced the impact of stripes and for a time cut down Maria's absences to about three an hour. Unfortunately this was not a long-term success, but Wilkins was sufficiently encouraged to continue with this work when he returned to the Medical Research Council Applied Psychology Unit in Cambridge.

There, he concentrated on a related condition, 'television epilepsy', in which people have seizures when watching TV. Television pictures are made up of many fine strips of flickering light, which could be the root of the problem. Wilkins found that television troubled people with epilepsy, migraine and dyslexia.

He devised a provocative pattern, drawn on paper, which was enough to give susceptible people headaches, eye-strain or seizures. It was a circle, 20 cms in diameter, filled with fine, horizontal, black and white stripes. It is unpleasant for anyone to look at, but for those prone to epilepsy or migraine, one glance can be enough to make them ill.

Research into the link between light and our health is relatively well established. Many find fluorescent strip lighting harsh and uncomfortable, but those prone to migraine or epilepsy are especially sensitive. Strobe lights in discotheques can trigger epileptic fits. Some people suffer from a severe depression in winter, seasonal affective disorder, caused by a lack of exposure to natural light. The standard, often successful, treatment is to sit under a light box every day, exposing yourself to particular wavelengths.

In agoraphobia, one survey found that more than a third of agoraphobics wear sunglasses when going out to try to reduce symptoms. Another found that a similar fraction prefer to go out in the dark – something Hippocrates noticed long ago. Almost three-quarters of the women in Dwarka's survey said they were especially sensitive to either natural or artificial lighting. Many particularly singled out the fluorescent lighting used in shops.

Wilkins also noticed that many agoraphobics complained that certain visual patterns caused them problems. A brief survey of agoraphobic volunteers found that more than one in four said strip lights caused 'very much' anxiety. Other troublesome lighting included glare, very bright sunshine and sunlight broken by trees or railings. All volunteers were asked to look at the provocative pattern and some saw illusions, such as colours, blurred or bending stripes. Those most sensitive to lighting were also most likely to have illusions when looking at the pattern.

When non-anxious controls took the same tests, Wilkins found that both groups were equally likely to get headaches when looking at the pattern or sitting in fluorescent lighting. However, only those with agoraphobia became anxious. A second survey among agoraphobics included visually stressful situations at home, such as strip lights in the kitchen. Even here, strip lights appeared to cause anxiety which suggested to Wilkins that the lighting alone could be responsible.

In offices, conventional fluorescent lights fluctuate in their intensity which can be unpleasant, even if you are unaware of the flicker. Wilkins found that half the headaches and eyestrain suffered by office workers could be put down to these fluctuations.

He examined people with agoraphobia under three different types of lighting: ordinary incandescent light, conventional fluorescent light and fluorescent light in which modern electronics reduces fluctuations. Under the conventional fluorescent light, which was indistinguishable from the adapted fluorescent light, agoraphobics had a higher heart rate and reported more illusions when they looked at the provocative pattern. Non-agoraphobics had many fewer illusions and symptoms. It suggests that the nervous system can respond to conventional fluorescent lights at a subconscious level.

People with epilepsy and those with agoraphobia react to the same sort of pattern, but this does not mean that the same thing is happening in their brains. Research discussed in chapter 4 suggests that panic attacks may be related to a discrete abnormality in part of the brain associated with emotion. Wilkins suggests that visual stimulation could somehow trigger a response in this part of the brain, but cannot say exactly how.

An alternative suggestion is that visual stimulation works via depersonalisation. This is an important part of agoraphobia in which people might describe themselves as feeling like a stranger to themselves, or as if they were separated from other people by a veil. Sometimes familiar things seem strange or their own body seems detached from themselves. It comes about when agoraphobics try to reduce their anxiety by actively taking in less information from the outside world. Unfortunately, this is a counterproductive strategy that can cause feelings of depersonalisation, which in turn leads to anxiety and depression. It is possible that the visual stimuli trigger the feelings of depersonalisation.

A third possibility is that the visual stimuli trigger anxiety only in agoraphobics because they interpret the unpleasant body sensations catastrophically. As discussed in chapter 6 it may be their interpretation of a minor bodily disturbance, their assumption that something is badly wrong, that sends them into panic.

This is extremely interesting research which is now lying dormant. Wilkins developed specially tinted glasses to reduce the pulsation from artificial lighting and found it very helpful to people with dyslexia. Wearing the glasses kept the lines on the page steady and people were able to read much more accurately. The idea that sunglasses help in agoraphobia could tie in here: people may have found an effective treatment for themselves. However, this assumption has yet to be properly tested.

## Alternative Applications

The circular wooden tub was almost five feet across and one foot deep. Inside, rows of bottles were arranged in circles, some pointing to the centre, others pointing outwards. It was filled with water, with iron filings and powdered glass sprinkled on liberally. The lid was closed and iron rods sticking out through it bent at right angles towards the edge of the tub.

A group standing round the tub each held on to an iron rod. They were deep in concentration, almost trance-like. They placed the rods against various parts of their bodies and held hands. This was treatment by 'animal magnetism' and it was used for a range of different illnesses.

It was devised by Franz Anton Mesmer, a doctor working in Paris in the late eighteenth century. He believed that invisible fluid flowed through the body according to the laws of magnetism. Any obstacles to the flow could cause disease, but the obstacles could be overcome by 'crises', the trance-like states which often ended in delirium or convulsions. Mesmer first treated high society, but his ideas became so popular that he devised ways of working with many people at once. The circular tub, or *baquet*, was one. Another was to 'magnetise' a tree and allow his patients to hold ropes hanging from its branches.

The medical profession of the day reacted violently against this theory. He was first drummed out of Vienna, accused of fraud, and moved to Paris. A few years later, King Louis XVI appointed a commission, including American statesman Benjamin Franklin and the French chemist Antoine-Laurent Lavoisier, to investigate his work. When they found no evidence for his claims, he was forced to leave the city.

Some of Mesmer's ideas about trances and hypnotism were most notably taken up and advanced by Sigmund Freud years later. Fortunately, his so-called forces of animal magnetism were largely left alone. But aspects of magnetism as medicine remain attractive even now. Mark George's work in chapter 4 is a testament to the possible uses of powerful and carefully directed electromagnetic fields; alternative practitioners are interested in using far lower levels. One gadget that has featured in magazines and newspapers, though not scientific journals, is 'Empulse', a magnetic field generator the size of a 50 pence piece. It is worn around the neck like a pendant and the manufacturer, MDI Ltd, claims that it works by emitting a series of pulsed magnetic frequencies that can boost specific activities in the body. Cell regeneration and growth, oxygen uptake, pain relief and muscular relaxation can all be encouraged by pulses of the right magnetic frequency, it says. The gadget is said to relieve the symptoms of a range of medical conditions including migraine, chronic fatigue and ME, multiple sclerosis, arthritic and back pain, allergies, insomnia, stress and tension.

The evidence for Empulse comes from satisfied customers, most of whom suffered from migraine. Many started having attacks as children and had previously tried everything that conventional medicine could offer without success. A doctor, for example, was able to play sport again and could work whole weekends on call with no problems. A woman who had avoided a host of foods for years suddenly had the luxury of becoming a chocoholic. Most convincing are the stories of patients who suddenly felt a return of their symptoms, only for them to discover later that the battery was flat in their Empulse.

Stories, however, are not scientific evidence. How many people used Empulse and did not improve at all? How do we know that these successes are not a tiny proportion of the total numbers given Empulse? The power of belief is so potent that some people got better holding on to ropes attached to Mesmer's magnetised tree. Only larger, carefully controlled studies can determine whether the device is truly having an effect. So far, the results of limited small-scale trials have been mildly encouraging, if short of revolutionary. MDI is now conducting a double-blind placebo-controlled trial of Empulse for migraine, and researchers in the US are testing its use in multiple sclerosis. The results will determine whether Empulse has a future in mainstream medicine.

Another device, which looks quite different, has inspired similar explanations. This is based on flickering lights and was initially an adapted pair of swimming goggles, devised by a neurologist in a London hospital. The goggles have now been replaced by the more sophisticated 'Lightmask', studded with flashing lights. They alternately illuminate the right and then left eye. You slip on the mask, close your eyes and sit back.

This work was originally directed at people with migraine, which was remarkable in that flashing lights are an important cause of the problem. But researchers found that if people wore the mask as soon as they had the first warning symptoms, they improved dramatically. Almost all migraines were lessened, more than half were prevented altogether. Like Empulse, Lightmask has been used to treat people with ME and has also been shown to reduce the anxiety related to PMS.

It is not known how the flashing lights help, but one explanation is that they, like Empulse, alter the frequency of electrical signals in the brain. Women with PMS have been shown to have an increase in the slow delta waves just before their period. This has also been associated with attention deficit disorder in children, chronic fatigue and minor head injuries. It is possible that Lightmask interacts with, and adjusts, these electrical signals.

Other explanations are that it could increase bloodflow to crucial parts of the brain, or that it works by resetting the internal circadian clock where biological rhythms are out of step with the normal sleep-wake cycle. Alternatively, it may exert its effect through relaxation; flickering lights, just like sitting by a real fire, might just make us feel better. Here, as elsewhere, many questions remain unanswered.

## Any Answers?

Mary Dwarka's story has opened up a whole new avenue of potential causes of phobias. Aspects of travel and transport are, by definition, difficult for agoraphobics but the idea that they could be solely responsible forces us to look more closely at light and electromagnetism.

There is enough evidence to conclude that types of artificial light and electromagnetism can have an effect on our general health, but so far only Wilkins has looked directly at agoraphobia. His finding, that strip lights can provoke anxiety in susceptible people, is one of the most compelling. Most studies to date provide tangential support, at best, for the idea of agoraphobia as a travel disorder.

Evolutionists are nowadays at pains to stress that evolution theory can be taken too far. Not every feature of human functioning can be explained with reference to the distant past. Sometimes we just are as we are, and the reasons are either recent or not at all clear. Do men really have a better spatial and navigational sense than women, or do we just like to credit them with it? Even supposing they do, is that because Stone Age men roamed further from home than the women who stayed at base? Or is it because many young boys today still receive more encouragement than girls do to play with cars and trains and are brought up with the image of themselves in the driving seat?

The disruption to navigation by women's hormones, found by Baker, is another case in point. He was unable to say whether premenstrual changes in women's hormones had a specific effect on orientation skills, or just made the women feel under par. Dwarka's own experience, of total freedom from agoraphobia while pregnant, again ties in neatly with the idea of women's vulnerability fluctuating at different times of their lives. But, as discussed in the last chapter, when the effects of pregnancy were studied systematically, the results were inconclusive. Some women got better during pregnancy, some got worse and some stayed the same. Despite the convincing stories, solid evidence to support this part of Dwarka's story does not exist.

The work explored in this chapter repeatedly suggests links between agoraphobia and a handful of other conditions. The zoologist, psychologist, GP and neurologist came from different standpoints to examine the effects of light and electromagnetism on health, but possible links between agoraphobia and dyslexia, women's hormones, migraine and ME popped up consistently throughout their work. Why should that be? One explanation is that agoraphobia is a complex condition and there are many different ways of becoming susceptible to it. Feelings of disorientation associated with dyslexia could contribute to agoraphobia. Anxiety associated with PMS or migraine could be another trigger, as could the lack of energy and depression in ME.

So – is agoraphobia a travel disorder? Many features of modern public transport can undoubtedly prompt agoraphobia: overcrowding, leading to feelings of claustrophobia and lack of an escape route; unpleasantly harsh strip lighting contributing to dizziness, headache and anxiety, perhaps even local electromagnetic fields disorientating us. But this could easily be only part of the problem, not the full explanation. None of the evidence is strong enough to say that agoraphobia is solely a travel disorder, and not anxiety-based. In a practical sense, it may not matter. Travel and transport are so difficult for people with agoraphobia that it could usefully be considered both a travel *and* an anxiety disorder.

In fact, Dwarka paints a rather depressing picture of the implications of a 'travel disorder' and her arguments may apply however agoraphobia is classified.

### The Potential for the Problem to Increase:
#### Changes in Travel
#### By Mary Dwarka

If we continue to alter our living conditions at the present rate, there is potential for certain categories of phobias, most noticeably agoraphobia, to increase. By contrast, the prevalence rate for animal and insect fears should not be expected to increase or decrease.

The pattern of travel which we have come to impose on ourselves in recent years bears no resemblance to the 'simple' navigation of our early ancestors. Environmental planners now talk of keeping Britain moving at the rate of 20,000 people an hour. We know very little about the psychological impact of these changes on us as individuals and whether we share a similar capacity to adjust.

Modern road schemes incorporating underground tunnels and elevated highways, for example, place travellers in confined settings. The heightened sense of restriction experienced by people with agoraphobia could be a first indicator that the disorder is likely to increase in the years ahead.

The ECA study showed that overall, public transportation was cited as the agoraphobia setting most avoided. The data suggest that a significant part of the population may be coping with their agoraphobia only because they have access to the use of a motor car or some other means of private transport. The findings are significant in that there is particular emphasis at present on discouraging car use. In Britain, there are plans to introduce extensive public transport schemes in the next two decades – such as hi-capacity rapid transit and computerised light-rail. This means that travelling, even within local communities, could become increasingly stressful for people who depend on public transport, and who are susceptible to agoraphobia. Again, it is women who may be marginalised. According to statistical data from the Department of

Transport, it is women in the general population who predominantly depend on public transport – compared with men, they make 50 per cent more journeys by bus, and 40 per cent more on foot.

At present, people coping with agoraphobia have to organise their employment and activities to take their circumstances into account, thus significantly reducing their capacity to be equal participants in society. Furthermore, the condition has a high level of stigma which means that individuals are tackling their difficulties virtually in isolation from the rest of the community. The psychiatrist Randolph Nesse believes that apart from the implications for research, the evolutionary theory for phobias has a more immediate impact – that once patients are aware that fears have their roots in their adaptive significance, this improves morale and their ability to tackle the problem. This theory could also make it possible for society as a whole to be more tolerant when addressing the issue of phobias.

# A Physical Problem?

## The Inner Ear

Forget all you have read so far. Forget genes, neurotransmitters and hormones. Forget habits in your behaviour, ways of thinking or anxious traits in your personality. Your phobia is in fact caused by a physical problem in your inner ear. Such defects account for nine out of ten phobias.

These extraordinary claims are being made by a New York professor. He says, furthermore, that medication aimed at correcting the defect produces dramatic and speedy improvements in most.

Harold Levinson has a solid medical background. He is clinical associate professor of psychiatry at New York Medical Center and runs a private practice in Great Neck, New York State. However, his offbeat approach to phobias and their treatment has pushed him well outside the mainstream.

Science has a habit of rewarding mavericks. The vast majority of researchers graft away in laboratories and clinics, testing out and extending existing theories. They painstakingly translate knowledge into treatments and push medicine forward, step by tiny step. They deserve our recognition and praise. However, real advances tend to come in giant leaps. They take us in unexpected directions and are frequently proposed by people under little obligation to follow the trends of mainstream thought. Any big new idea threatens the status quo of the prevailing scientific establishment. New ideas are not always welcome.

Of course, new ideas are not always right, either. Progress in science depends on a combination of the maverick with a bright idea, and the grafter with the patience and determination to see if it is valid.

Much of this book has been devoted to current research and the potential it holds for the future. Some looks into our nature, the part played by genes, neurotransmitters and hormones, our gender perhaps. Some looks into the importance of nurturing, and includes the environment in the womb, our parents' attitudes, our upbringing and how we see the world. Even the possible influence of external factors like light and electromagnetism is being investigated.

But what if Levinson is right and phobias are really a lot simpler? What if their root cause is physical and one that can almost always be successfully treated? If true, this would certainly revolutionise phobia research.

Levinson's theory is based around a part of the ear and brain which has nothing to do with our hearing. Sound is processed by the outer and middle ear. The inner ear lies beside the middle ear but works in tandem with an ancient part of the brain called the cerebellum. Together, they control balance and coordination and process sensory information. The inner ear is comprised of three semi-circular canals at right angles to one another, lined with tiny hairs. When we move our head, fluid in the canals also moves and bends the hairs. This sets up a signal which is sent to the cerebellum for interpretation and allows us to orientate ourselves, to balance and coordinate.

Any defect in the system can lead to dizziness and the nausea associated with motion sickness. It can leave someone feeling clumsy or bewildered, overwhelmed by the movement in crowds, as if suffering from a sensory overload. According to Levinson, these sensations can become the basis of a whole range of phobias. Fear of heights might be related to dizziness. Agoraphobia could be linked with an inability to cope with crowds. Fear of travelling by car or plane could be caused by motion sickness. Social phobia could originate in poor co-ordination and a tendency to spill drinks or knock things over.

The explanation sounds reasonable enough so far but what makes Levinson so controversial is the extremity of his claims and the lack of hard data to support them. Virtually nothing in medicine can explain such a wide range of symptoms. No psychiatric treatment has this level of success. Levinson's claim to have accounted for 90 per cent of phobias is so high that it is unbelievable to most doctors, and it would take some well-designed and extremely large studies to satisfy them. He has published papers on the neurological tests described later, designed to pick up problems in the inner ear/cerebellar system. People with anxiety disorders performed significantly worse than others in these tests, suggesting a link between anxiety and the inner ear. However, even he concluded that further, larger studies by other groups would be necessary to confirm it.

Others have been reluctant to take up the challenge and he is far from conciliatory. His books claim there has been a conspiracy to cover up his work and subtitles such as 'A Scientific Watergate' do not endear him to other researchers.

Levinson is viewed as such an outsider that doctors have refused to discuss his work for fear of lending it undue credibility.

But we should pause before dismissing Levinson's claims out of hand. The success rate and the numbers he quotes sound too good to be true, but what of the theory itself? If he is correct for half, a quarter or even a tiny fraction of those with phobias, his work would be a valuable contribution. He has amassed convincing stories of the benefits of his approach and plenty of phobics can relate readily to some of his descriptions. They might have always felt different from their friends. They may not have started out as particularly anxious but always had terrible coordination, been clumsy and unable to play sport at school. They may have had intermittent dizzy spells and always hated flying or travelling by car.

Many of Levinson's patients are attracted by things they have read in the media and they are a so-called self-selected group. Those whose symptoms most nearly fit the model he describes are most likely to be inspired to go to him for treatment. So nine out of ten of *his* patients could, possibly, have problems with their inner ear. Others who find the explanation less convincing will be much less likely to be attracted to his work.

The very idea of a physical cause for phobias pushes the discussion into semantics. Phobias are partially defined as being irrational fears. If the fear is reasonable because of an underlying physical cause, then it is not strictly a phobia. The *DSM* specifically excludes fear relating to physical illness. For example, someone with Parkinson's disease who has shaky hands and becomes reluctant to eat or drink in public for fear of rattling a plate or spilling tea does not have a true phobia. The fear is not classed as social phobia, but rather as a secondary problem of Parkinson's disease, even though the distress and avoidance may be exactly the same.

A whole list of physical conditions are linked with the

symptoms of panic. Low blood sugar levels, perhaps due to mild or undiagnosed diabetes, can cause faintness. Benign postural vertigo makes you feel dizzy if you stand up too quickly. Acute inflammation of the inner ear, or labyrinthitis, can trigger dizziness or vertigo. Some prescription drugs can increase metabolism and heart rate in a way similar to caffeine.

The root cause of conditions like diabetes or labyrinthitis requires treatment, first and foremost. This should relieve uncomfortable feelings of dizziness, palpitations or vertigo and reduce the symptoms of panic. But there may be a second tier of physical problems usually so mild that they go unnoticed for years on end, sometimes for life. Treatment may not be appropriate. It may not even exist. But these conditions could have a subtle link with phobias, and agoraphobia in particular.

## Palpitations

Another New York psychiatrist, Dr Charlotte Zitrin, suggested twenty years ago that phobias, and agoraphobia in particular, could be related to physical problems. Some of her work touches on Levinson's, but her claims are more moderate and research-based. In work that was partly echoed more recently by Sydney Brandon in the UK (see chapter 4), she pointed out that agoraphobics can be grouped according to their most troublesome symptoms. Some are particularly afraid of cardiac arrest and often describe their wild heartbeat. Others primarily have respiratory symptoms. They feel as if they are going to suffocate when they have a panic attack, they become breathless, start hyperventilating and make themselves dizzy. Others have gastro-intestinal symptoms. They have nausea, feel sure they are about to vomit or have to rush to the toilet with sudden diarrhoea. She suggested that these different groups of patients might in fact have different underlying physical causes.

Zitrin was one of the first to formally investigate the relationship between agoraphobia and a syndrome affecting the heart. The mitral valve is the gateway separating two chambers of the heart. It opens to allow blood to flow into the heart and closes as the heart beats and pumps blood around the body. A problem called mitral valve prolapse syndrome is very common and normally so minor that it frequently goes unnoticed for the whole of someone's life. However, those with this syndrome often get palpitations and it can be picked up by doctors monitoring the heart.

Cardiologists have found that about one in twenty people with no symptoms at all have mitral valve prolapse. In Britain alone, then, it could affect two or three million people. Researchers who have studied the syndrome have concluded that this may not be as alarming as it sounds. Only half experience palpitations and only a fraction are troubled by them.

Zitrin joined forces with cardiologists at Long Island Jewish-Hillside Medical Center to examine the possible relationship between mitral valve prolapse and agoraphobia. They looked at twenty-five women with agoraphobia who complained especially of palpitations and compared these with twenty-three others who had no history either of heart disease or panic attacks. The women's hearts were monitored objectively and the researchers found that eleven of those with agoraphobia had evidence of mitral valve prolapse. Only two of the women in the control group showed any signs of it.

This meant that the women with agoraphobia were five times more likely to have mitral valve prolapse than those with no anxiety disorder and suggested that a heart syndrome could be involved in the development of panic attacks and agoraphobia. Zitrin did not study exactly how palpitations lead to agoraphobia but her assumption ties in with the 'vicious circles' discussed in chapter 6. Someone with mitral valve prolapse may have frequent palpitations. If they are afraid of the sensation, they may become aroused with fear, which in turn will make the palpitations worse. This vicious circle can spark off a panic attack. The panic attack can be such a terrible experience that people start to avoid situations in which they think it may occur. Their avoidance grows and eventually leads to panic attacks and agoraphobia.

Zitrin and her colleagues concluded that there is probably a whole spectrum of reactions to the palpitations. At one extreme is a total lack of concern. In the middle are those who react to the palpitations and are disturbed by them. Their fear increases and instigates panic attacks, yet they do not avoid situations in which they have had the attacks. Despite regular panic attacks, they never develop agoraphobia. At the other end of the spectrum are those so distressed by the sensations that they avoid any situation likely to bring on a bout of palpitations. Over time, the number of places they avoid increases and they develop full-blown agoraphobia.

It is the reaction to the palpitations, rather than the palpitations themselves, that leads to agoraphobia. Palpitations alone cannot be sufficient to cause agoraphobia since millions of people in the US and Europe have palpitations and few develop agoraphobia. Palpitations have to be combined with another factor, probably psychological. Nevertheless, the palpitations are real and may have a physical cause. Many who have suffered with agoraphobia for years might never have developed it had they not also had a mitral valve prolapse in their heart. This syndrome, which is physical and can be measured, was the first step in their path towards agoraphobia.

Equally, many could be free from anxiety disorders simply because they have no recurrent physical symptoms. They may possess the psychological background that can turn a symptom into agoraphobia, but without the symptoms they are lucky enough to escape the possible consequences.

Many agoraphobics are suspicious of psychological approaches because they are certain that the initial stage in their panic attack, the palpitations, is real, and not triggered by stress. It can be extremely frustrating for someone who is confident they are not being over-emotional to be asked to focus on their reactions rather than the sensations. Pinning down the physical cause of the symptoms can be helpful. Naming the syndrome gives confidence that the medical world not only believes in it, but has seen it before and found it benign. Simple reassurance that palpitations are unlikely to do any damage can be enough to tone down reactions and to nip problems in the bud. Once the fear has been quelled, palpitations are less likely to escalate into a panic attack.

People with well-established agoraphobia can also find such advice a turning point. Careful reassurance that convinces them the palpitations are not harmful can help them break the vicious circle that keeps the agoraphobia in place. Once they are sure that a panic attack cannot kill, it makes less sense for them to avoid the places and situations they have come to fear. Again, this reasoning echoes that discussed in chapter 6.

Mitral valve prolapse is not the only medical syndrome to cause palpitations. An overactive thyroid gland produces hormones that speed up the metabolism. This problem, called hyperthyroidism, causes weight loss because the body starts to burn up fat stores more vigorously than usual. Sufferers may may also lose muscle mass and become weak. At the same time, the heart rate can speed up and they may be prone to palpitations. Zitrin gathered some evidence that hyperthyroidism, like mitral valve prolapse, might be associated with agoraphobia, presumably for similar reasons.

Physical problems relate to agoraphobia and not to any other phobia, according to Zitrin. In a lifetime's work, she found no link with either specific phobias or social phobias. This may be because the physical problem, and the interpretation of it, escalate in a vicious circle to produce panic attacks. Panic attacks are a core feature only of agoraphobia, not of specific or social phobias.

Her work on mitral valve prolapse and agoraphobia has been replicated by other groups in other settings, essential for being taken seriously by mainstream science. However, the *DSM* says the evidence remains equivocal. Some groups found that people who had panic attacks were more likely to have mitral valve prolapse or thyroid disease than the general population. Other groups, however, found no difference. So the jury is still out on whether there is a true association between such physical problems and agoraphobia, but Zitrin and others believe that a range of physical symptoms may be capable of providing the initial prompt that can eventually lead to agoraphobia.

Another group, whose panic attacks involve mainly respiratory symptoms – gasping, breathlessness, dizziness and feeling faint – may have lactate imbalance (see chapter 4). Their panic attacks sound quite different from those with palpitations as the primary problem. But people with these two distinct starting points who share the psychological characteristics necessary to turn a physical symptom into agoraphobia could eventually end up with similar problems.

Others with gastrointestinal symptoms could have irritable bowel syndrome, for instance. Anyone terrified of suddenly needing the toilet is unlikely to be happy sitting in the middle of a long row at the theatre. If this fear escalates to the point where they start avoiding theatres, restaurants, etc., they could be taking their first steps towards agoraphobia.

Another group of agoraphobics have vertigo as their main symptom. They feel dizzy looking down from bridges, hate escalators and tall office blocks. They would not consider climbing the Eiffel Tower on a trip to Paris or even going up a small church tower to look at the view. Zitrin again has unpublished data that a condition called labyrinthitis may sometimes be related to these symptoms. Labyrinthitis is an inflammation of the inner ear, or labyrinth. So, like Levinson after her, Zitrin examined the inner ear for a link with phobias and believes it probably exists. Her claims for its involvement, though, are much more modest than his.

Motion sickness and vertigo can make us acutely aware of our inner ear and cerebellum. A rough Channel crossing can overstimulate the system and produce nausea, sweating, pallor and vomiting. Vertigo makes us feel as if we are moving when we are actually quite still.

If one inner ear is damaged, through physical injury or infection, the results can be dramatic. A rat with only one inner ear will roll over and over as it tries to stand up. People with a damaged inner ear show less obvious problems but can suffer acutely. Even if they lie down and keep absolutely still they are likely to suffer waves of nausea, vomiting, sometimes even diarrhoea. They feel every bit as bad as the passenger on the rough ferry crossing and do not have the luxury of being able to get off the boat when it reaches dry land. Nerves from the damaged inner ear keep sending a false message to the cerebellum that the person is moving about. Fortunately, even if the inner ear does not recover, the body can compensate. The cerebellum stops responding to the false messages and after a couple of months the symptoms often disappear completely.

This is the extreme case in which the inner ear is permanently damaged. But more transient problems can also have long-term knock-on effects. Another set of Zitrin's patients had inflammation of the labyrinths and it gave them vertigo. She contends that some reacted so badly that it set them on the road to agoraphobia. People with mitral valve prolapse syndrome were so afraid of the palpitations that their fear made the symptoms worse and led them to develop panic attacks and agoraphobia. In the same way, those with vertigo, she says, may have been so distressed by the unpleasant symptoms that their anxiety made the symptoms worse and they spiralled into a panic attack. Both groups could end up with similar avoidance patterns and agoraphobic symptoms, even though they started off with two quite different physical problems.

Her evidence is far from conclusive and these assumptions preliminary. But it echoes work that is sufficiently serious to be referred to, if not endorsed, in the *DSM*. The link with the seasickness study mentioned in chapter 8 is also interesting.

Researchers simply noted that seasickness, like agoraphobia, is more common in women. Zitrin's work tentatively suggests that an underlying reason for both conditions could rest in a damaged inner ear.

Levinson's and Zitrin's work overlaps only in as far as they agree that the dizziness so often associated with panic attacks and agoraphobia may be linked to physical defects in our balance mechanism. However, Zitrin's statements were cautious and she insisted that psychological factors must also be present before agoraphobia can develop. By contrast, Levinson's enthusiasm for the idea of a physical cause is apparently boundless. Like Klein (see chapter 4), he insists that the psychological and emotional element of phobias has been overplayed.

## Dyslexia and a Lack of Balance

Levinson, like Zitrin, worked at Long Island Jewish-Hillside Medical Center, but he came to his theory on the physical cause of phobias in a different and roundabout way. He started out as a Freudian-inspired psychiatrist and psychoanalyst, and has devoted much of his working life to phobias and anxiety disorders. As he was building up his private practice, he worked with dyslexic children in schools in New York in order to support himself. Dyslexia is generally considered a disorder in which people scramble or reverse letters, numbers and words and, not surprisingly, often have severe reading difficulties. Levinson examined 1,000 children with dyslexia, looking initially for evidence of problems in the thinking brain, low IQ or problems in comprehension and production of speech, for example.

He did not find this. Instead, as he says in his book *Phobia Free*, he found evidence of balance and coordination problems in three-quarters of the children. In a further study with other dyslexics, he says 95 per cent, or nineteen out of twenty people, had these signs of dysfunction in the inner ear and cerebellum. When this was checked out in a hospital setting by other clinicians, the dysfunction was found in nine out of ten.

A link between dyslexia and the inner ear is perhaps even less intuitive than the link with phobias. But other researchers have similar findings. A group in Sheffield were thinking along the same lines and developed some odd-looking tests to measure young children's balance. Researchers blindfolded children and asked them to stand with one foot in front of the other, their arms stretched out in front. They were videotaped and the researchers measured the size and frequency of children's wobbles.

They discovered that children who wobble a lot and seem to have poor balance are also likely to have dyslexia. Research is on-going and the Sheffield group is developing a range of tests to pick up problems. They have used brain scans to show that dyslexic adults have abnormalities in the cerebellum. Eventually, they hope to test children starting school and pick out those who are likely to have trouble reading. This could spare children the years of anguish they often have to endure before their dyslexia is diagnosed.

The Sheffield research is based on the premise that the cerebellum is involved in thinking and language processing, as well as posture, balance and coordination. Measuring wobbles picks up balance problems which might signal difficulties in processing language, since the same part of the brain handles both functions.

Levinson, however, says that dyslexia is not a reading or language disorder, but rather a coordination problem. People's eyes cannot track the written word across the page, which is why it can help to have fewer or bigger words. If dyslexia were a reading problem, in which people did not understand the symbolic significance of a letter, the size of the print would make no difference.

This reasoning fits in with Arnold Wilkins's work, discussed in the previous chapter. Wilkins studied the effect of fluorescent lighting on both dyslexia and agoraphobia. One of his many findings was that dyslexic children's reading improved, sometimes dramatically, if they wore specially tinted glasses. The research comes from a different angle, but it suggests that some aspects of dyslexia are related to the mechanics of reading, to the ease with which the eye can move across the words on a page. A child's ability to interpret letters and words may be flawless, but the words appear in a haphazard order, it is not surprising that they don't make sense. Both Wilkins and Levinson believe that dyslexia could be a mechanical rather than a comprehension problem. Levinson, however, believes dogmatically that the problem lies in the cerebellar-vestibular system.

Whatever the underlying theory, Levinson reasoned that if this system is involved with dyslexia, drugs that are useful in treating other cerebellar-vestibular problems, such as motion sickness, might also be helpful in dyslexia. So he started prescribing drugs such as antihistamines for children with dyslexia. He hoped for an improvement in their reading ability and he found it. But what he also saw, he says, was an improvement in all sorts of different symptoms. Children became more confident, less anxious, panicky and moody. Those plagued by obsessive, repetitive thoughts and actions said they felt free of these feelings, often for the first time.

Children with dyslexia often have phobias, says Levinson. It is easy to see why a child who is falling behind in their reading may start to dislike classes and develop a phobia of school. But according to Levinson they might also have one or more phobias because of their inner ear problems.

Levinson classifies phobias on the basis of their assumed cause. Motion-related phobias include fear of lifts and escalators as well as of boats, aircraft or cars. Balance-related phobias boil down to problems with the inner ear as gyroscope and include fear of heights, bridges and stairs. Coordination-related phobias make people afraid of driving if they are unable to steer a car along a straight line or to make judgements when travelling at speed. In more extreme cases, coordination phobias can include fear of eating in public, fear of sports, dancing and sex, as all involve movement. Compass-related phobias are those in which sufferers feel disorientated, lose their sense of direction and their ability to judge distance with any accuracy. This description is reminiscent of Mary Dwarka's thesis in the previous chapter, and according to Levinson, these phobias may be made worse by tunnels and enclosed spaces, open spaces and darkness. They include fear of crowds, supermarkets and shopping malls.

When Levinson's children were treated for dyslexia, their phobias disappeared or improved markedly. He says this was totally unexpected and initially attributed it to the surge of good feeling among children who have had so many of their frustrations eliminated. You might expect that a child who has had problems with memory and concentration, has been hyperactive and easily distracted, would receive a tremendous boost after a sudden improvement. Under these circumstances, it is not surprising that their anxiety would plummet and their phobias melt away.

But Levinson claims there is more to it than that. He believes dyslexia and phobias are both a direct result of the same inner ear or cerebellar dysfunction. Where he had previously assumed he was treating two quite separate groups of patients, dyslexics and phobics, he came to suspect that dyslexia and phobias are two different manifestations of the same underlying disorder.

He then went back to patients with phobias and asked them directly about possible dyslexic symptoms. Many revealed problems with memory, concentration, reading, writing and spelling. Almost all had a history of inner ear symptoms, he says, and had neurological evidence of dysfunction in this system.

He now uses similar neurological tests to diagnose both dyslexia and phobias because he is examining the function of the inner ear in both cases. The tests include an eye-tracking test in which a pen is moved from side to side a few inches in front of a patient's face. Most can follow the movement with their eyes, but when the pen is stopped suddenly, those with inner ear dysfunction can be identified because their eyeballs continue to move back and forth. Other tests sound more like the Sheffield researchers'. People are asked to walk in a straight line, putting one foot in front of the other, heel to toe. Swaying and loss of balance are typical signs of inner ear dysfunction. Or they may be asked to shut their eyes, hold their arms in front of them and lift one leg. Irregular movements of the eyes are another giveaway sign.

Levinson claims that more than one in five of the general population have some degree of inner ear disturbance. Some have a genetically determined problem. Others acquire it in the womb because their mother had an infection during pregnancy, fell or was malnourished. A premature or difficult birth – involving improper use of forceps, or oxygen deprivation after the umbilical cord became wrapped around the baby's neck – could damage a previously healthy inner ear system. A range of infections in childhood or adolescence, including ear or sinus infections and glandular fever, can interfere with its development, as can drug use, surgery, anaesthesia, whiplash injury, malnutrition, prolonged emotional stress, hormonal imbalances during the menopause or the degenerative changes associated with old age. It is not difficult, according to Levinson, to acquire an inner ear dysfunction.

They are also straightforward to treat, he says, primarily with antihistamine and seasickness pills which work directly on the faulty inner ear. Some are available over the counter, but he stresses that nobody should use them for treating phobias unless under the supervision of their doctor.

Levinson's medication list gives his critics some of their most potent ammunition. Many of the antihistamine and seasickness treatments he names also have a sedative or anti-anxiety effect. Many drugs for motion sickness come with warnings that people should not drive or operate machinery after taking them. Critics claim that the drugs he suggests may indeed reduce panic, but this could be as a side effect of their main action. The effect of the drugs on the inner ear system may be irrelevant.

Worse, Levinson also has a list of secondary medications, which include compounds absolutely familiar to any psychiatrist. They include the benzodiazepines and imipramine, which have been discussed intermittently throughout this book. They have been standard treatments for anxiety and depression respectively. So, whatever the strengths and weaknesses of Levinson's theory, his treatment may work because it includes tried and tested anti-anxiety drugs.

It is possible, though, that his explanation may be helpful in itself. Even today, psychological and psychiatric disorders carry such a stigma that a diagnosis of a physical disorder is often preferable, particularly when it describes a patient's experience so accurately. Levinson gives a physical diagnosis, and then prescribes the same treatment as many more conventional psychiatrists.

Levinson counters that such criticism misses the main points. These diverse groups of drugs have all been shown to have effects on the inner ear system, or on motion sickness. This is not a chance finding, he says, but is further validation of the connection between inner ear dysfunction, anxiety and phobias, dyslexia, balance, coordination and so on.

For yet another possibility is that Levinson is right, at least for a proportion of phobias. Maybe dysfunction of the inner ear really can lead to phobias, just as Zitrin's other physical symptoms may play a role.

The idea that agoraphobia comes in many different forms is now accepted. Many scientists are convinced that there are different subgroups of agoraphobia. The single label of agoraphobia may exist for no better reason than that we are not sure how to classify types usefully. One future approach might be to classify the type of agoraphobia according to the type of symptoms. Those who mainly suffer palpitations could be separated from those whose main problem is breathlessness and suffocation. Even though the final condition, agoraphobia, may have identical effects on both subgroups, this re-classification could help researchers systematically explore possible starting points for agoraphobia.

In the end, as so aptly demonstrated by Levinson, the most effective treatment for phobias may not depend on the precise cause. He recommends psychotherapy for his patients, he says, because many have developed poor self-esteem and lack of confidence over years without a proper diagnosis or treatment. Doctors agree then, that people with phobias benefit from cognitive behavioural therapy or medication, or some combination of these, and the reasons why they became phobic in the first place may play little part in the choice of treatment. But many sufferers who are convinced that they had a physical problem in the first place can be reassured and encouraged that the part played by physical illness is now being seriously considered.

# Conclusion

## Into the Future

Fifty or a hundred years from now, it is tempting to wonder where all this research, imagination, thought and work will have led. Will phobias have been eliminated? Will our descendants look back at the beginning of the twenty-first century and wonder at the unnecessary misery endured by so many? Or will anxiety and phobias have reached such endemic proportions that they will view our lives with wistful envy, longing for the freedom from angst most of us now enjoy?

Our knowledge of the mechanics of anxiety, fear and phobias will only increase. Neuroscience is at such an exciting stage it is inconceivable that we will not have many more facts at our fingertips over the next few years. Even politicians – who are not generally renowned for their interest in science – have been convinced of its progress. In October 1989, then President George Bush announced that the 1990s were to be the 'Decade of the Brain', dedicated to the encouragement and redoubling of research efforts into its workings and its many disorders.

The initiative was welcomed by neuroscientists and psychiatrists, well used to working in the underfunded services of modern healthcare. Their specialities had finally come of age. Our knowledge of the brain may still be embryonic compared with that of the rest of the body, but at least it is now considered a serious science.

Neuroscientists have a daunting subject: the brain is perhaps the most complex structure on earth. It weighs about three pounds on average but is made up of a hundred billion or more nerve cells, each able to connect with thousands of other cells. Signals are sent at speeds of more than 200 miles an hour and have to be precisely coordinated for even the simplest task to be accomplished. At the beginning of the 1980s, only a handful of neurotransmitters had been identified. By the end there were more than forty. In 1980, the study of genetic disorders had barely started; by the end of the decade scientists had pinpointed the genes for specific disorders and started looking at genetically linked diseases. Replacement or repair of damaged neurones was just a pipe dream but by 1989 animal research had demonstrated that nerves can re-grow. New scanning technology, which allows researchers to examine the brain while people are alive and conscious, became available. Previously, brains could only be studied after death.

Progress accelerated through the 1990s. Further advances in imaging allowed clinical researchers to observe the living brain in exquisite detail. Fundamental discoveries at cell level hinted at how the brain translates chemistry into behaviour, thought and emotion. Molecular biology and genetics started to revolutionise the study of inheritable neurological disorders. New understanding of the biochemistry of neurotransmitters meant that, for the first time, drugs could be designed to act on specific defects with minimal side effects. Previously, drugs had been discovered by trial and error. Put together, this work started to reveal the brain's structure and how it affects our development, health and behaviour.

The organisers of a meeting in Washington in 1999 declared, perhaps over-excitedly, that the decade had 'delivered more advances than all previous years of neuroscience research combined'. None the less, the progress was stunning and much of it relevant in the search for an improved understanding of anxiety, panic and phobias.

President Bush's announcement was not an offer of funding, rather a call for increased awareness. But it was certainly driven by financial considerations. Congress's resolution on the decade noted that fifty million Americans are affected each year by brain disorders and disabilities. These include the major mental illnesses; inherited and degenerative diseases; stroke; epilepsy; addictive disorders; prenatal injury, environmental toxins and trauma; and speech, language, hearing and cognitive disorders. In 1989, total annual treatment, rehabilitation and related costs came to an estimated US $305 billion.

The announcement was also politically appealing because it backed a near-certain winner. There were guaranteed to be advances. The politicians' confidence was rewarded, and by the end of the decade even notoriously cautious scientists started to sound optimistic. An editorial in *Nature Neuroscience* noted that the research community has increasing confidence that 'brain disease is now a tractable problem'. Disorders that were 'once considered inevitable features of the human condition, are now seen as specific diseases whose causes can be identified and which it will some day be possible to prevent or cure'. An article in the *Lancet* added, 'Psychiatry has lost its aura of hopelessness.' Scientists and clinicians alike believe they are standing on the edge of a new era of understanding. But what exactly does that mean for some-body with a phobia?

## The Next Hundred Years

Unfortunately, understanding does not necessarily wipe out a problem. It would be wonderful to think that total knowledge equals total resolution but in medicine this is rarely the case. Many common and serious diseases are well understood. Scientists have tracked some of them down to the level of the individual cell and gained understanding vital in the long-term battle to find cures. But this understanding cannot always prevent problems occurring in the first place. In asthma, for example, expanding knowledge about what is going on in the cells lining the tiny passages in the lungs has been translated into new, more effective drugs. But the improvements are in treatment not prevention. Asthma is increasingly common in the Western world with more children suffering and more dying from it. Presumably some change in our way of living is to blame, but nobody knows exactly what. It has been variously suggested that houses are too clean or too dirty, too damp, dusty or full of house mites. Increased pollution could be the culprit, a sedentary lifestyle, even immunisations or artificial additives to our food. New information about the workings of lung cells has improved treatment, but the broader questions of why so many children have asthma and how it could be prevented have gone unanswered.

Phobias, likewise, will become more effectively treated. The fields of research discussed in this book are already producing better drugs and more efficient talking cures. Some, such as the study of temperament and personality, are hinting at ways of preventing the development of phobias in individual children at risk. Future generations of scientists will build on the work of today's geneticists, neurologists, cognitive and behavioural therapists. Treatments in the future might be hormonal, psychological or pharmaceutical, or even involve highly specific magnets and electric currents. They will be rigorously assessed and increasingly imaginative and effective.

Not all current lines of research will be fruitful – some will no doubt fall from fashion or reach dead ends without producing useful implications. Others will translate into progress. Santiago Ramón y Cagal's discovery of the nerve cell a century ago set the scene for much of today's research in the neurosciences, but the work of many of his contemporaries disappeared without trace. Some of today's work is laying foundations for the science of the future but it would take courage or recklessness to identify the theories most likely to pave the way.

We can at least be confident that treatments will improve. Whether this will help to reduce the incidence of phobias in those at risk, or in society at large, is another question. One hundred years from now, phobias could be far more or less common than they are today. Does current research tell us anything about what to expect in the future? Strictly speaking, no. The theories covered in this book attempt to explain what is happening now, a difficult enough task in itself. Most are concerned with the process within an individual with a phobia, while a few, such as evolution theory, look at phobias in society at large. None of the work discussed sets out to predict the future. However, if we allow ourselves a large measure of scientific licence and extrapolate from the research covered here, it might be possible to pick out potential trends for the future.

It is always humbling for scientists to look back through history at the real advances in human health. Many can be traced to improvements in standards of living rather than detailed knowledge of disease processes. Underground sewers and the provision of clean water have probably saved more lives than any other single innovation. Better housing and healthier diets have had more impact on women's health, and therefore on infant mortality, than any new obstetric practice. The old high school debate of whether a dustman should be paid more than a doctor hinges on the truth that households benefit more from having their waste taken away than from anything a doctor can do.

Thus it follows that the future prevalence of phobias could depend heavily on changes to our general living environment, living standards and features of our everyday life. Changes in lifestyle might be as important as any of the minutiae of scientific progress. And this century is set to see big shifts in the way we live.

The advent of the silicon chip has sparked a time of extraordinary technological change which is affecting all aspects of our daily routines, work and play. Over the next one hundred years, computers will change our lifestyles radically and alter our vulnerability to fears and anxiety. The computer revolution could be directly responsible for an increase or decrease in the numbers of those who fall prey to phobias.

The primary effect of improving communications is on our working lives. Computers connect up the globe. Electronic mail allows us to communicate effectively across continents. Satellites can link all employees of an organisation wherever they are located, whether office, sitting room, golf course or beach. Video conferencing means that staff in London, New York and Tokyo can have face-to-face meetings without leaving their own cities, as long as some are prepared to work at an unsociable hour. All this is starting to make our traditional work patterns less relevant. Employers may no longer want staff to work in office premises, five days a week. Staff may prefer not to travel to the office but work instead from home, perhaps for several different employers, or clients. The home office is a reality and part-time, self-employed and so-called portfolio working are, increasingly, the norm.

A workforce based in homes across the country might be an extremely satisfactory arrangement all round. Many businesses will be able to cut the costs of maintaining city-centre premises. Skilled staff and contractors will eliminate hours of travelling from their weekly schedules. People working in all sorts of different industries will no longer have to travel to work every day – no more waiting for buses or trains, or sitting in traffic jams. Furthermore, new work patterns will allow us greater choice in where we live. We could choose to live in remote and beautiful parts of the world, perhaps in small towns and villages, becoming part of a local community rather than an anonymous fragment of city life.

This all sounds like good news, but the side effects of these changes for anyone vulnerable to phobias are complex. Work is an important part of our lives and we devote a large portion of our time to it. These new ways of working will have the dual effect of both widening and narrowing our horizons. On the one hand, electronic communication opens up our world. We can have easy, daily contact with distant colleagues. On the other hand, a working life conducted from our own sitting room can be an isolating experience. Colleagues provide stimulation and chatter, and are often the focus of our social life as well. A workforce based in individual homes scattered many miles apart may be more convenient, but less fun. Such changes are likely to be helpful to some people vulnerable to phobias, and decidedly unhelpful to others.

Evolution theory concerns itself with the broader picture of why phobias are so common. Evolution does not move at any great pace and the theory assumes that our most basic instincts are little different from those of our ancestors living 10,000 years ago. Phobias arise because of a mismatch between today's environment and that of the Stone Age, which means we suspect danger where none exists. Systems that warn us of dangers in our environment can be too finely tuned for modern life and alarm bells ring too easily.

Our instincts will not change appreciably over the next hundred years, but if technology increases the pace of change, which seems likely, our descendants could be even more out of tune than we are. Hunter-gatherers lived before the introduction of agriculture and were forced to devote most of their time to the search for food. They would starve if they could not find enough to eat. It is possible that a group that had encountered a predatory animal would stay at base for a few days to allow the danger to pass, but most people, most of the time, would be out looking for food.

The modern equivalent is the daily commute to work. Roaming the land searching for berries might seem easy set against fighting your way on to a train at rush hour. Possibly it was, a lot of the time. But the Stone-Ager was at real risk from animals and might sensibly stay at base when something dangerous was prowling nearby. Evolutionists say that agoraphobia is an exaggerated but natural response to threat; many of those with agoraphobia recognise that they face no real danger, no matter how much they hate the journey.

The removal of the daily commute in years to come could have a complex impact on agoraphobia. People who cannot face the journey to work sometimes have to give up good jobs. Work practices that encourage working from home could allow them to stay. This may not address the problem of agoraphobia directly, but the boost to self-esteem associated with carrying out satisfying work should not be underestimated. This would obviously be a positive change.

On the other hand, there may be many more who are vulnerable to agoraphobia than actually suffer from it. These people may dislike or even dread travelling to work but somehow, day after day, they succeed in making the journey. Combining the thoughts of evolutionists and behaviourists, this daily struggle may well be a good thing. Behaviourists say that people are helped simply by carrying out the task they dread, no matter what their thoughts about it or their reasons for doing so. However unpleasant the daily journey seems, it might be a healthier alternative for those susceptible to agoraphobia than sitting at home all day with no need to step outside at all. The enforced routine of leaving the house every morning and travelling to work might be the bane of someone's life, but if it forces them to get out and meet others it could be protecting them from the much worse problems associated with developing full-blown agoraphobia.

This ties in with the observation that many women develop agoraphobia after they leave busy jobs to spend time looking after children. Becoming a parent is a major upheaval, and a multitude of factors could set the scene for the development of agoraphobia. Women's hormones may go haywire during pregnancy, sleep deprivation is a tried and tested form of torture and a reduced disposable income can put a heavy strain on relationships. A contracted social life is not the only important contributory factor, but the isolation often associated with motherhood has been singled out as a potential trigger for the development of a debilitating phobia. Working from home may not effect hormones or sleep patterns but by removing people from a busy environment to work alone it could predispose some to developing agoraphobia.

Similar arguments also apply to social phobia. Stone-Agers lived in small groups and interacted within their own well-defined, familiar circle. Their instincts, like ours, were honed to allow them to interpret tiny cues in others' facial expressions, body language or tone of voice. They had to know their place within a group because they could be expelled for upsetting its leader. As survival depended on group cooperation, this would probably mean death.

Even now, at work and in our social life, we judge instinctively how highly we are regarded. Our instincts are often more accurate than our ears and we usually have a sense that we are falling out of favour long before anyone tells us. Our boss may praise a piece of work but if the praise is not genuine our antennae, developed over tens of thousands of years, will be twitching.

Electronic communications, at a stroke, could make these instincts almost redundant. The global workplace of the twenty-first century means that we will be dealing with, and depending on, people we have never actually met. Talking to clients from a distance via technology will camouflage tell-tale signs of trouble. Communicating with a colleague on a video link will convey only a fraction of the implicit information transmitted in a face-to-face meeting. Anyone who has worked in television will tell you that, contrary to the popular saying, the camera *does* lie. Larger than life characters shrink when conveyed on video. Equally, people who appear extremely natural and friendly on screen are not necessarily approachable off camera. An accomplished performer can totally hide their feelings. Electronic communications that mask signs of minor disapproval are doing no favours if they prevent us from rescuing a situation before it has gone too far. The consequences of falling out with a boss or client might not be as drastic as in the Stone Age – we are unlikely to die as a result – but social standing and financial security could be on the line.

However, once again the impact of this technology is complex. Anyone with social phobia could find electronic communications a godsend. Not having to meet a client's eye, not having to go for a drink or meal with them, could remove all the most dreaded aspects of certain occupations. It could enable social phobics to hold down positions they would otherwise have been unable to contemplate, and to progress in prestigious, well-paid careers. It may not address the underlying anxiety, but it will prevent the anxiety setting up a chain of events which wreak havoc on their lives.

On the other hand, those prone to the phobia will no longer have to face the kind of daily behaviour therapy which office life can offer. Those who hate the social side of their work, but participate anyway, may be extremely relieved to have a way out. But in the long term, if this new way of working means they never have to face their fears, phobias may develop unchecked. Changes in working practices may help those with entrenched social phobia, like those with agoraphobia but, again, may tip the balance for many others with fears only just under control.

Solitary working might also have adverse effects on minor anxieties, if we follow the principle of vicarious learning encompassed within evolution theory. As discussed in chapter 2, almost every monkey reared in the wild is afraid of snakes. Not every one of these monkeys can have encountered a snake, and Sue Mineka's work demonstrated that monkeys learn fear after watching the reactions of a dominant monkey. Similarly, they can be immunised against developing fear by watching other monkeys failing to respond to a snake.

Without wanting to take the analogy too far, our immediate social environment colours our view of how dangerous the world is. We learn from our own experiences but also from those of esteemed friends and colleagues. Numerous studies into crime have found that we tend to think the world is more dangerous than it actually is, and we often overestimate our chances of being mugged as we walk along the street. The more we socialise, at home and at work, the more normal and positive experiences we accumulate and the more immunised we are against developing fears. Estimations of danger remain estimations, and do not progress into fear. Isolating a workforce removes these normalising experiences and could allow caution to develop into unrealistic anxiety.

This may be an unnecessarily pessimistic view of the future. A home workforce need not be isolated. People who are unhappy living in huge cities may be well suited to life in smaller communities. They would lose the society of workmates but, having exchanged anonymity in a city for life at the heart of a village, they will have a new, very different social scene. Knowing all one's neighbours is rare in a large town, but more normal in villages – and more like the social set-up of the Stone Age. We might be well adapted to living in communities where everyone knows everyone else. In this way, technological advances could bring society round full circle.

Another important discrepancy exists between prehistoric life and life in the future: speed and distance of travel has changed out of all recognition. Hunter-gatherers travelled on foot and returned to their home base at least every few nights. Their natural in-built navigation aids and use of the earth's electromagnetic field, discussed in chapter 9, reminded them constantly of where they were in relation to home. Modern transport, with its artificial lighting, electromagnetic fields, huge speeds and distances, may confuse this natural ability to navigate and could make vulnerable people feel disorientated and uneasy. We spend hours of each day in artificial light, which may be to blame for some migraine and depression as well as for agoraphobia. Future studies will pinpoint the type of lighting, or even electromagnetism, that causes problems, and contribute to the design of newer and more comfortable public transport systems.

Damage to the inner ear was implicated as a cause of phobias in chapter 10. It can be caused by trauma at birth or later infections. If it is responsible for some phobias, both should occur less frequently in future as general medical procedures improve. If the inner ear does have a central role to play, we can expect to see a reduction in phobias as the years pass.

The changing role of women in the Western world has been the cultural revolution of the past few decades and could be a cornerstone in the reduction of agoraphobia. Women are assuming aspects of men's lives unthinkable a generation ago. Career prospects are improving and their earning power is increasing. The flexible working practices of the computer age may accelerate these changes if they suit women better than existing office hierarchies. Discrimination on the grounds of sex will be less likely in a world in which contractors at the end of a satellite link are hired for specific skills. Changes like this may be happening too slowly for many women, but in evolutionary terms they have come about in the blink of an eye.

Improved earnings for women are having a knock-on effect in the home. The career-minded are less likely to stop working when they have children. Many households rely on two incomes and though househusbands are still rare, it often does make economic sense for the woman to be the main breadwinner. Shifts in attitudes towards women are filtering down and affecting the way children are treated. Society still expects different things from girls and boys, but there is more cross-over now than there has ever been. We have always expected girls to be sociable and to have domestic skills, but now we want them to achieve academically at school and in sport as well.

These changes could be important for the future prevalence of phobias. Women in the late twenty-first century will be far removed from Stone Age culture, but this could be to their advantage. Phobias, especially agoraphobia, are so much more common in women that a change which makes their lives more like men's have traditionally been, could protect from phobias.

There are, of course, many possible causes of phobias. Social changes will have no impact on hormone levels and women are always going to be biologically distinct from men. But behaviourists and cognitivists agree that the way we are taught to think and behave is a key to our vulnerability to phobias. Traditional social roles could have been predisposing women to developing phobias. Attitudes that have prevented girls from exploring rough play may have exaggerated underlying feelings of physical vulnerability. Encouraging girls to stay near home, or praising them for being tidy, neat and clean could make the outside world seem dangerous or just best avoided. If these attitudes lose their grip over the next few decades, and young girls are allowed to have broader experience and aspirations, phobias will develop less often.

Overall, changes in lifestyle that could potentially generate more phobias over the next hundred years may be offset to a healthy degree by parallel changes in social attitudes. Our basic predisposition to phobias could remain pretty much unchanged if we assume that lifestyle is isolated from medical advances. But that, of course, is unlikely.

## Breaking Down Barriers

One of the most exciting moves in current phobia and anxiety research is the breaking down of barriers between specialities. Scientists, philosophers, psychologists and psychiatrists approach the problem from different angles, choose different landmarks and describe it in different language. Their viewpoints complement and build on each another and success in any one field should benefit all others.

Science has historically gained from cross-fertilisation of ideas between specialities. Ideas which are long accepted within one field can be revolutionary in another. As mentioned in chapter 1, the engineers who invented the pump hundreds of years ago provided the background for the breakthrough in understanding the heart and circulation. Anatomists and physiologists could only compare the heart to a pump after engineers had invented it. Similarly, developments in computer technology have helped neurophysiologists understand how parts of the brain work. In turn, computer specialists have drawn from evolutionary principles to produce algorithms by which programs improve themselves over time.

Disciplines relating to phobias have shown little interest in each other for too long. They have been separated by invisible walls of ignorance, suspicion and lack of respect. When behaviourist Isaac Marks teamed up with US psychiatrist Randolph Nesse to write up their evolutionary perspective on phobias, they struggled to find a publisher. Journals keen to publish their other work were sceptical about this apparent departure into another field. Within five years, evolution theory has gained so much ground that Marks and Nesse would have no problem if they were to submit a similar paper today. However, the reluctance of journals to publish new areas of research is telling.

The first cracks in the barriers between disciplines are appearing and could herald a new era in research. Theories in this book may be widely differing but in the main they are not in conflict. Although they may contradict in details, the broad sweep of each is compatible.

Evolution, which looks at *why* we have phobias rather than how they develop, provides a framework for thinking through the other theories. It is almost a philosophy rather than a science and yet it leads directly into one of the most highly advanced fields of research today. Genetics and evolution go hand in hand. If fear has survived intact down countless generations, keeping our ancestors alive, it could be written into our genes.

Geneticists today are unravelling the code and a hundred years from now the implications of this research will be everywhere. Chapter 3 outlined the problems encountered in the search for panic, anxiety and phobia genes, many of which will have been overcome by the year 2100. We could have a full understanding of the mechanisms involved in such disorders, along with a new range of highly specific treatments.

These treatments may eventually work directly on our genes, modulating their effects, but so far gene therapy has had limited success. Even in comparatively straightforward genetic disorders like cystic fibrosis, scientists have struggled for years to replace a single gene. Undoubtedly it will be far harder to regulate the activity of the many genes likely to influence phobias.

Long before then, scientists will be working on the downstream effects of genes, in our chemical and hormonal make-up. Genetics research translates directly into neurophysiology. Genes regulate the production of chemicals in our brains and influence the sympathetic nervous system, determining our heart rate and how easily we are aroused by frightening events. They regulate our stress and sex hormones, which may in turn affect our vulnerability to phobias.

Once genes are found, the next step is to determine exactly how they work and whether drugs can be designed to intervene. Many current drugs were discovered by chance and have a range of wanted and unwanted effects throughout the body. In future, drugs will be designed to reach specific molecular targets where they will have the desired effect only, with no side effects.

Equally, learned behaviour also fits the evolutionary model. Anxiety is passed down the generations by learning, each individual looking to their parents for guidance on how to behave in novel circumstances. Most of us learn appropriate levels of fear and anxiety and in turn pass this on to our children. But if, for example, a child has a terrifying experience and is left with excessive or inappropriate fear, he or she may think and act differently forever afterwards.

Behavioural and cognitive therapists try to get us to unlearn unhelpful habits. Neither bother unduly with how the fear was acquired but instead aim to teach ways of beating it. Behaviourists focus on the way we act when we are afraid, and cognitive therapists aim to change the way we think. Together the therapies counteract fear, whether it was learned, or caused by genes, neurochemicals or hormones.

The entirely positive developments to spring from genetics are only marred by the question of how genetic information could be used and misused. Genetic information is qualitatively different from other medical information in that it is an individual's blueprint, the template from which their body and all its systems develop. In future, print-outs of an individual's genetic make-up may be available from the moment of birth, or even of conception. The most likely causes of death may be known before a baby is even born.

Genetic research, therefore, is conducted within an ethical minefield. Long before gene therapy becomes routine, many diseases could be prevented by genetic investigation during pregnancy, but only at the terrible cost of terminating affected foetuses. Eugenics has cast a long shadow and the suspicion that scientists are trying to create designer babies is fuelled rather than extinguished by reports of progress. Scientists' claim that the primary effect of research is to improve our understanding of diseases and develop better treatments for them is lost against the idea that only those unlikely to develop the diseases will ever be born.

However, anxiety, panic and phobias are neither fatal nor simple and are unlikely to take centre stage in ethical debates on genetics. They involve a complex interaction between many different genes and the environment. When anxiety genes are isolated, and it becomes routine for us all to know the salient points of our genetic make-up, parents may be aware, even before birth, that their child is susceptible to developing anxiety. They could then be helped to create the right sort of environment for their child. This would apply to all sorts of illnesses. Just as some seeds require certain types of soil in order to grow and flower, some children need a particular home environment if they are to develop their full potential. Genes for cancers and heart disease might encourage healthy diets and exercise; anxiety genes might encourage parents to adapt a naturally cautious child's environment to halt the development of debilitating anxiety disorders. In this future world, of course, any phobias or anxiety disorders suffered by parents would have been successfully treated, thus eliminating one of the child's risk factors from the outset.

The counterargument is that common-sense and careful observation already allow parents to spot an anxious temperament without the need for futuristic innovations. We know that genes influence our neurophysiology and hormonal make-up, and contribute to our personality and temperament, but so what? Chapter 7 mentioned the tiny baby who was so highly strung that her parents had to remove the mobile hung over her cot because it over-stimulated her. Her parents did not need to know the details of her genetic make-up to understand that she was nervous. Many parents automatically adapt to the personality and temperament of their children and implement *ad hoc* strategies to help them through difficult times. In this way, implementation of today's knowledge could provide good solutions much sooner than waiting for the genetic code to be cracked.

But high-tech work might be the only way of chipping away at the mystery surrounding phobias. Few geneticists searching for a root cause of anxiety would say that their primary aim was to remove the stigma associated with phobias, but it could be an important side effect. Parents who blame themselves can be reassured that their child's nervousness is not their fault, nor that of the home environment they provided. Similarly, many adults with phobias are ashamed of their apparent weakness. If they knew that their fear had a biological cause, they might be more inclined to seek help. The current conspiracy of silence around phobias creates untold unnecessary misery and if genetics contributes nothing more than to make society more sympathetic to these disorders, it will have carried out a vital role. If a fear of buttons or a horror of wigs is known to be caused in part by a chemical reaction in the brain, then the public at large may finally be able to regard phobias as no more embarrassing than swollen tonsils.

It is ironic that an important result of highly sophisticated genetic research could be a surge in demand for low technology specialists in cognitive and behaviour therapy. But it demonstrates again how the different disciplines are starting to interact. High-tech research is allowing these disciplines to prove themselves. As neurotransmitters can be measured precisely, and the meanings of the levels known, or as imaging techniques allow scientists to watch our brains as we laugh, cry or panic, then the least data-driven of specialities will be able to demonstrate their effectiveness in an evidence-based world.

Just as taking drugs affects behaviour, psychological interventions alter brain chemical levels. Mind and body are coming together. An innovative project at the University of California used yoga to treat obsessive-compulsive disorders. Researchers showed that it was as effective as drugs in reducing symptoms, whereas general relaxation and meditation had no effect. Their next step is to scan the brain and see exactly how, and where, yoga works.

Research like this is terrifically encouraging at a time when the molecular sciences were in danger of tearing psychiatry and psychology apart. An understanding of the brain will require the different skills and viewpoints of many specialities. High-tech science cannot produce that understanding alone. Diseases may originate in genes or neurotransmitters, but they take hold and grow in the mind. To look at the cause without considering the expression is to see only half the story.

What this means for the late twenty-first century and beyond is that phobias are set to be transformed from vaguely embarrassing and apparently irrational fears into a series of specific disorders with well-known causes. The stigma that still surrounds phobias could finally evaporate.

A range of much-improved treatments will be available. Drugs will work quickly and have few side effects. Talking treatments will have been honed down to the bare essentials. Behavioural therapy will be tailored to the individual so that fears are addressed directly. Psychoanalysis will still exist and may be more highly valued in future because its special skill of finding meaning in emotion will help us make sense of the mass of molecular data.

A full range of treatments will be available and anyone embarking on treatment will be advised in advance of the approaches most likely to succeed for them. But prevention will be the norm. Children at high risk will be picked up early, even before birth, and appropriate interventions put in place.

None of this research aims to eliminate natural caution, nobody wants to reduce the variety of personalities and temperaments that so enrich our lives. But the crippling fear and anxiety, and the sheer misery that phobias can bring, do little to enhance our world and I imagine that in a hundred years' time, our descendants will be happy never to witness or have to experience them.

# Further Reading

## Introduction

### Fight the Fear

Robins, L. N., Helzer, J. E., Weissman, M. M., et al., 'Lifetime prevalence of specific psychiatric disorders in three sites', *Arch. Gen. Psychiatry* (1984), 41: 949–58.

Klerman, G. L., Hirschfeld, R. M. A., Weissman, M. M., et al. (eds), *Panic Anxiety and its Treatments: Report of the World Psychiatric Association Presidential Educational Task Force*, Washington, DC, American Psychiatric Press Inc. (1993), ISBN: 0-88048-684-8.

### From Antophobia to Zoophobia

Canino, G. J., Bird, H. R., Shrout, P. E., et al., 'The prevalence of specific psychiatric disorders in Puerto Rico', *Arch. Gen. Psychiatry* (1987), 44: 727–35.

Morakinyo, O., 'Phobic states presenting as somatic complaints syndromes in Nigeria: socio-cultural factors associated with diagnosis and psychotherapy', *Acta. Psychiatr. Scand.* (1985), 71: 356–65.

*Diagnostic and Statistical Manual of Mental Disorders*, 4th edn, Washington, DC, American Psychiatric Association (1994).

Furmark, T., Tillfors, M., Everz, P., et al., 'Social phobia in the general population: prevalence and sociodemographic profile', *Soc. Psychiatry Psychiatr. Epidemiol.* (1999), 34 (8): 416–24.

## Chapter 1: History

### In the Beginning

Thorpe, S. J., Salkovskis, P. M., chapter 4, 'Animal Phobias', in *Phobias: A Handbook of Theory, Research and Treatment*, (ed.) G. C. L. Davey Chichester, John Wiley & Sons Ltd (1997). ISBN: 0-47196-983-4.

Sims, A., 'Historical aspects of anxiety', *Postgraduate Medical Journal* (1988), 64 (suppl. 2): 3–9.

Westphal, C., 'Die Agoraphobie: Eine neuropathische Erscheinung', *Arch. Psychiatrie Nervenkrankheiten* (1871), 3: 138–61.

McNab, B. *Perceptions of Phobia and Phobics*, Academic Press, London (1993). ISBN: 0-12485-960-7.

### Cartesian Logic

Marías, J., *History of Philosophy*, New York, Dover Publications Inc. (1967), ISBN: 486-21739-6.

### Locke and Empiricism

Locke, J., *Some Thoughts Concerning Education*, London, Ward Lock (1693): 481–2.

### Darwin and the Dawn of Modern Science

Richards, G., *Putting Psychology in its Place: An Introduction from a Critical Historical Perspective*, London, Routledge (1996), ISBN: 0-41512-863-3.

Darwin, C., *The Origin of Species*, reprinted in Penguin Classics (1985), ISBN: 0-14043-205-1.

## Freud on Fear

Freud, S., *Collected Papers*, vol. 3, London, Hogarth Press (1950).

Wolpe, J., Rachman, S., 'Psychoanalytic "evidence": a critique based on Freud's case of Little Hans', *Journal of Nervous and Mental Disease* (1960), 130: 135–48.

Holmes, J., 'The assault on Freud', *Current Opinion in Psychiatry* (1996), 9: 175–6.

Frances, A., Miele, G. M., Widiger, T. A., et al. 'The classification of panic disorders: from Freud to DSM-IV', *J. Psychiat. Res.* (1993), 27 (suppl. 1): 3–10.

## Little Albert

Watson, J. B., Rayner, P., 'Conditioned emotional responses', *J. Exp. Psychol.* (1920), 3: 1–14.

Watson, J. B., 'Psychology as the behaviorist views it', *Psychological Review* (1913), 20: 158–77.

Pavlov, I. P., *Conditioned Reflexes*, transl. G. V. Anrep, London, Oxford University Press (1927).

## Chapter 2: Evolution

## Living Without Fear

Darwin, C., *The Origin of Species*, reprinted in Penguin Classics (1985), ISBN: 0-14043-205-1.

Marks, I. M., Nesse, R. M., 'Fear and fitness: an evolutionary analysis of anxiety disorders', *Ethology and Sociobiology* (1994), 15: 247–61.

Nesse, R. M., 'An evolutionary perspective on panic disorder and agoraphobia', *Ethology and Sociobiology* (1987), 8: 73S–83S.

Darwin, C., *The Expression of the Emotions in Man and Animals*, (Third Edition), (ed.) Paul Ekman, London, Fontana Press (1999), ISBN: 0-00638-734-9.

Davey, G. C. L., 'Preparedness and Phobias: Specific Evolved Associations or a Generalized Expectancy Bias?' *Behavioral and*

*Brain Sciences*, Cambridge University Press (1995), 18 (2): 289–325.

## Age-Old Anxieties

Poulton, R., Davies, S., Menzies, R. G., Langley, J. D., Silva, P. A., 'Evidence for a non-associative model of the acquisition of a fear of heights'. *Behav. Res. Ther.* (1998), 36 (5): 537–44.

## Animal Instincts

Raguram, R., Bhide, A. V., 'Patterns of phobic neurosis: a retrospective study', *British Journal of Psychiatry* (1985), 147: 557–60.

Zafiropoulou, M., McPherson, F. M., ' "Preparedness" and the severity and outcome of clinical phobia', *Behav. Res. Ther.* (1986), 24 (2): 221–2.

de Silva, P., 'Phobias and preparedness: replication and extension', *Behav. Res. Ther.* (1988), 26 (1): 97–8.

Seligman, M. E. P., 'Phobias and preparedness', *Behavior Therapy* (1971), 2: 307–20.

Mineka, S., Davidson, M., Cook, M., et al., 'Observational conditioning of snake fear in rhesus monkeys', *Journal of Abnormal Psychology* (1984), 93 (4): 355–72.

Cook, M., Mineka, S., Wolkenstein, B., et al., 'Observational conditioning of snake fear in unrelated rhesus monkeys', *Journal of Abnormal Psychology* (1985), 94 (4): 591–610.

Cook, M., Mineka, S., 'Observational conditioning of fear to fear relevant versus fear-irrelevant stimuli in rhesus monkeys', *Journal of Abnormal Psychology* (1989), 98 (4): 448–59.

Bennett-Levy, J., Marteau, T., 'Fear of animals: what is prepared?' *British Journal of Psychology* (1984), 75: 37–42.

Merckelbach, H., van den Hout, M. A., van der Molen, M., 'Fear of animals: correlations between fear ratings and perceived characteristics', *Psychological Reports* (1987), 60: 1203–9.

Öhman, A., Dimberg, U., Ost, L-G., 'Animal and social phobias: biological constraints on learned fear responses', in *Theoretical*

*Issues in Behaviour Therapy*, S. Reiss and R. R. Bootzin (eds), New York, Academic Press (1985), 107–33. ISBN: 0-12586-360-8.

Öhman, A., Soares, J. J. F., 'On the automatic nature of phobic fear: conditioned electrodermal responses to masked fear-relevant stimuli', *Journal of Abnormal Psychology* (1993), 102 (1): 121–32.

Cook, E. W. III, Hodes, R. L., Lang, P. J., 'Preparedness and phobia; effects of stimulus content on human visceral conditioning', *Journal of Abnormal Psychology* (1986), 95 (3): 195–207.

McNally, R. J., Foa, E. B., 'Preparedness and resistance to extinction to fear-relevant stimuli: a failure to replicate', *Behav. Res. Ther.* (1986), 24 (5): 529–35.

## Chapter 3: Genetics

### Icy Fear

Kagan, J., *Galen's Prophecy: Temperament in Human Nature*, London, Basic Books (1994), ISBN: 1-85343-390-X.

### A Family Story

*Diagnostic and Statistical Manual of Mental Disorders*, 4th edn, Washington, DC, American Psychiatric Association (1994).

Weissman, M. M., 'Family genetic studies of panic disorder', *J. Psychiatr. Res.* (1993), 27 (suppl. 1): 69–78.

Moran, C., Andrews, G., 'The familial occurrence of agoraphobia', *British Journal of Psychiatry* (1985), 146: 262–7.

Ost, L. G., 'Blood and injection phobia: background and cognitive, physiological and behavioural variables', *J. Abnormal. Psychol.* (1992), 101: 68–74.

### Bundles and Blankets

Fyer, A. J., Manuzza, S., et al., 'Familial transmission of simple phobias and fears: a preliminary report', *Arch. Gen. Psych.* (1990), 47: 252.

Fyer, A. J., Manuzza, S., Chapman, T. F., et al., 'A direct interview family study of social phobia', *Arch. Gen. Psych.* (1993), 50: 286–93.


## Double Trouble

Skre, I., Onstad, S., Torgeson, S., et al., 'A twin study of DSM-III-R anxiety disorders', *Acta. Psychiatri. Scand.* (1993), 88: 85–92.
Martin, N. G., Jardine, R., Andrews, G., Heath, A. C., 'Anxiety disorders and neuroticism: are there genetic factors specific to panic?' *Acta. Psychiatri. Scand.* (1988), 77: 698–706.
Andrews, G., Stewart, G., Allen, R., Henderson, A. S., 'The genetics of six neurotic disorders: a twin study', *J. Affective Disord.* (1990), 19: 23–9.


## Computer Synthesis

Kendler, K. S., Neale, M. C., Kessler, R. C., Heath, A. C., Eaves, L. J., 'Major depression and phobias: the genetic and environmental sources of comorbidity', *Psychological Medicine* (1993), 23: 361–71.


## The Hare and the Tortoise

Crowe, R. R., Noyes, R., Pauls, D. L., Slymen, D., 'A family study of panic disorder', *Archives of General Psychiatry* (1983), 40: 1065–9.
Harris, E. L., Noyes, R., Crowe, R. R., Chaudry, D. R., 'Family study of agoraphobia', *Archives of General Psychiatry* (1983), 40: 1061–4.
Noyes, R. Jr, Crowe, R. R., Harris, E. L., Hamra, B. J., McChesney, C. M., Chaudry, D. R., 'Relationship between panic disorder and agoraphobia', *Archives of General Psychiatry* (1986), 43: 227–32.
Crowe, R. R., Noyes, R. Jr, Wilson, A. F., Elston, R. C., Ward, L. J., 'A linkage study of panic disorder', *Archives of General Psychiatry* (1987), 44: 933–7.
Flint, J., Corley, R., DeFries, J. C., et al., 'A simple genetic basis for a complex psychological trait in laboratory mice', *Science* (1995), 269: 1432–5.

## Chapter 4: Neurophysiology

### Personal Magnetism

George, M. S., Wassermann, E. M., Post, R. M., 'Transcranial magnetic stimulation: a neuropsychiatric tool for the twenty-first century', *Journal of Neuropsychiatry and Clinical Neurosciences* (1996), 8: 373–82.

George, M. S., Speer, A. M., Bohning, D. E., et al., 'New methods for understanding how the brain regulates mood: serial perfusion fMRI and transcranial magnetic stimulation', in *Psychiatric neuro-imaging research: contemporary strategies*, Rauch and Dougherty (eds.), American Psychiatric Association Press, Washington, DC (2001).

### Hearts and Minds (and Lungs)

DaCosta, J. M., 'On irritable heart; a clinical study of a form of functional cardiac disorder and its consequences', *American Journal of Medicinal Science* (1871), 61: 17–52.

Drury, A. N., 'The percentage of carbon dioxide in the alveolar air and the tolerance to accumulating carbon dioxide in cases of so-called "irritable heart" of soldiers', *Heart* (London, 1919), 7: 165–73.

Pitts, F. M., McClure, J. N., 'Lactate metabolism in anxiety neurosis', *New England Journal of Medicine* (1967), 277: 1329–36.

### A Breath of Fresh Air

Klein, D. F., 'False suffocation alarms, spontaneous panics and related conditions: an integrative hypothesis', *Arch. Gen. Psych.* (1993), 50 (4): 306–17.

Klein, D. F., 'Testing the suffocation false alarm theory of panic disorder', *Anxiety* (1994), 1 (1): 1–7.

Pine, D. S., Weese-Mayer, D. E., Silvestri, J. M., et al., 'Anxiety and congenital central hypoventilation syndrome', *Am. J. Psychiatry* (1994), 151: 864–70.

Breslau, N., Klein, D. F., 'Smoking and panic attacks: an epidemiological investigation', *Arch. Gen. Psychiatry* (1999), 56 (12): 1141–7.

Briggs, A. C., Stretch, D. D., Brandon, S., 'Subtyping of panic disorder by symptom profile', *Br. J. Psychiatry* (1993), 163: 201–9.

## Ready for Anything

Holden, A. E., Barlow, D. G., 'Heart rate and heart rate variability recorded in vivo in agoraphobics and nonphobics', *Behavior Therapy* (1986), 17: 26–42.

## Adrenaline

Nutt, D., Lawson, C., 'Panic attacks: a neurochemical overview of models and mechanisms' (1992), 160: 165–78.

Charney, D. S., Heninger, G. R., Breier, A., 'Noradrenergic function and panic anxiety effects of yohimbine in healthy subjects and patients with agoraphobic and panic disorder', *Arch. Gen. Psychiatry* (1984), 41: 751–63.

Redmond, D. E. Jr, Huang, Y. H., 'Current concepts II: new evidence for a locus-coeruleus-norepinephrine connection with anxiety', *Life Sciences* (1979), 25 (26): 2149–62.

Uhde, T., Stein, M. B., Vittone, B. J., et al., 'Behavioral and physiologic effects of short-term and long-term administration of clonidine in panic disorder', *Arch. Gen. Psychiatry* (1989), 46: 170–77.

## Sleep No More

Mellman, T. A., Uhde, T. W., 'Sleep panic attacks: new clinical findings and theoretical implications', *Am. J. Psychiatry* (1989), 146 (9): 1204–7.

## Caffeine High

Bernstein, G. A., Carroll, M. E., Crosby, R. D., et al., 'Caffeine effects on learning, performance and anxiety in normal school-age children', *J. Am. Acad. Child Adolesc. Psychiatry* (1994), 33 (3): 407–15.

## States and Traits

Uhde, T. W., 'Anxiety and growth disturbance: is there a connection? A review of biological studies in social phobia', *J. Clin. Psychiatry* (1994), 55 (6, suppl.): 17–27.

Uhde, T. W., Tancer, M. E., Rubinow, D. R., et al., 'Evidence for hypothalamo-growth-hormone dysfunction in panic disorder: profile of growth hormone (GH) responses to clonidine, yohimbine, caffeine, glucose, GRF and TRH in panic disorder patients versus healthy volunteers', *Neuropsychopharmacology* (1992), 6 (2): 101–18.

Uhde, T. W., 'Growth hormone dysregulation in nervous humans and animals: practical and theoretical implications', chapter in *Growth Stature and Adaptation: Behavioral, Social and Cognitive Aspects of Growth Delay*, Stabler, B. and Underwood, L. E. (eds.), The University of North Carolina, Chapel Hill (1994), 83–98.

## Chapter 5: Behaviour

## Self Help

Ghosh, A., Marks, I. M., 'Self-treatment of agoraphobia by exposure', *Behaviour Therapy* (1987), 18: 3–16.

Ghosh, A., Marks, I. M., Carr, A. C., 'Therapist contact and outcome of self-exposure treatment for phobias: a controlled study', *British Journal of Psychiatry* (1988), 152: 234–8.

## Fight Fear With Fear

Rachman, S., Lopatka, C., 'A simple method for determining the functional independence of two or more fears', *Behaviour Research and Therapy* (1986), 24: 661–4.

## Behaviour Therapy – the Easy Way?

Marks, I. M., Swinson, R. P., Basoglu, M., et al., 'Alprazolam and exposure alone and combined in panic disorder with agoraphobia. A controlled study in London and Toronto', *Br. J. Psychiatry* (1993), 162: 776–87.

Heimberg, R. G., Liebowitz, M. R., Hope, D. A., et al., 'Cognitive behavioral group therapy vs phenelzine therapy for social phobia: 12-week outcome', *Arch. Gen. Psychiatry* (1998), 55 (12): 1133–41.

Liebowitz, M. R., Heimberg, R. G., Schneider, F. R., et al., 'Cognitive-behavioral group therapy versus phenelzine in social phobia: long term outcome', *Depress. Anxiety* (1999), 10 (3): 89–98.

## Courage Under Fire

Mowrer, O. H., 'Stimulus response theory of anxiety', *Psychological Review* (1939), 46: 553–65.

Rachman, S. J., *Fear and Courage*, 2nd edn, New York, W. H. Freeman and Co. (1990), ISBN: 0-71672-061-2.

---

## Chapter 6: Cognition

---

## Testing

Beck, A. T., *Cognitive Therapy and the Emotional Disorders*, New York, International Universities Press (1976). ISBN: 0-45200-928-6.

Beck, A. T., Emery, G., Greenberg, R. L., *Anxiety Disorders and Phobias*, New York, Basic Books (1985). ISBN: 0-46500-384-2.

Clark, D. M., 'Cognitive therapy for anxiety', *Behavioural Psychotherapy* (1986), 14: 283–94.

Clark, D. M., Beck, A. T., 'Cognitive approaches', in *Handbook of Anxiety Disorders*, C. G. Last and M. Hersen (eds), New York, Pergamon (1988). ISBN: 0-00032-766-4.

Salkovskis, P. M., 'The Cognitive Approach to Anxiety: Threat Beliefs, Safety-Seeking Behavior, and the Special Case of Health Anxiety and Obsessions.' Chapter in P. M. Salkovskis (ed.), *Frontiers of Cognitive Therapy*. New York, The Guilford Press (1996). ISBN: 1-5723-012-0.

Salkovskis, P. M., 'The importance of behaviour in the maintenance of anxiety and panic: a cognitive account', *Behavioural Psychotherapy* (1991), 19: 6–19.

## Disgust

Davey, G. C. L., 'Self-reported fears to common indigenous animals in an adult UK population: the role of disgust sensitivity', *British Journal of Psychology* (1994), 85: 541–4.

Webb, K., Davey, G. C. L., 'Disgust sensitivity and fear of animals: effect of exposure to violent or revulsive material', *Anxiety, Stress and Coping* (1992), 5: 329–35.

Rozin, P., Fallon, A. E., 'A perspective on disgust', *Psychological Review* (1987), 94: 23–41.

Davey, C. L. G., 'Disgust', in V. S. Ramchandran (ed.), *Encyclopaedia of Human Behavior*, vol. 2, San Diego, Academic Press Inc. (1994), 135–41. ISBN: 0-12226-920-9.

Mulkens, S. A., de Jong, P. J., Merckelbach, H., 'Disgust and spider phobia', *J. Abnormal Psychology* (1996), 105 (3): 464–8.

Thorpe, S. J., Salkovskis, P. M., 'Studies on the role of disgust in the acquisition and maintenance of specific phobias', *Behav. Res. Ther.* (1998), 36 (9): 877–93.

## Risk Assessment

Tomarken, A. J., Mineka, S., Cook, M., 'Fear-relevant selective associations and co-variation bias', *Journal of Abnormal Psychology* (1989), 98: 381–94.

Mathews, A., MacLeod, C., 'Selective processing of threat cues in anxiety states', *Behaviour Research and Therapy* (1985), 23: 563–9.

Nesse, R. M., Klaas, R., 'Risk perception by patients with anxiety disorders', *The Journal of Nervous and Mental Disease* (1994), 182: 465–70.

Mathews, A., MacLeod, C., 'Discrimination of threat cues without awareness in anxiety states', *Journal of Abnormal Psychology* (1996), 95 (2): 131–8.

## Chapter 7: Personality and Temperament

### Three-Dimensional Personalities

Eysenck, H. J., *The Structure of Human Personality*, London, Methuen (1970). ISBN: 0-41618-030-2.

Cloninger, C. R., Przybeck, T. R., Svrakic, D. M., The tridimensional personality questionnaire: US normative data', *Psychol. Rep.* (1991), 69 (3, pt 1): 1047–57.

Brown, S. L., Svrakic, D. M., Przybeck, T. R., Cloninger, C. R., 'The relationship of personality to mood and anxiety states: a dimensional approach', *J. Psychiatr. Research* (1992), 26 (3): 197–211.

Sigvardsson, S., Bohman, M., von Knorring, A-L., Cloninger, C. R., Symptom patterns and causes of somatization in men: I differentiation of two discrete disorders', *Genetic Epidemiology* (1986), 3: 153–69.

### The Inner Child

Kagan, J., *Galen's Prophecy*, New York, Basic Books (1994): ISBN: 1-85343-390-X.

Garcia-Coll, C., Kagan, J., Reznick, J. S., 'Behavioural inhibition in young children', *Child Development* (1984), 55: 1005–9.

Kagan, J., Reznick, J. S., Snidman, N., 'Biological bases of childhood shyness', *Science* (1988), 240: 167–71.

Kagan, J., Snidman, N., Arcus, D., 'Childhood derivatives of high and low reactivity in infancy', *Child Dev.* (1998), 69 (6): 1483–93. *New Scientist*, 22 June 1996.

## Family Likeness

Murray, L., Stanley, C., Hooper, R., King, F., Fiori-Cowley, A., 'The role of infant factors in postnatal depression and mother–infant interactions', *Dev. Med. Child Neurol.* (1996), 38 (2): 109–19.

## Inhibition, Fear and Phobias

Rosenbaum, J. F., Biederman, J., Gersten, M., et al., 'Behavioral inhibition in children of parents with panic disorder and agoraphobia: a controlled study', *Arch. Gen. Psychiatry* (1988), 45 (5): 463–70.

Rosenbaum, J. F., Biederman, J., Bolduc-Murphy, E. A., et al., 'Comorbidity of parental anxiety disorders at risk for childhood-onset anxiety in inhibited children', *Am. J. Psychiatry* (1992), 149 (4): 475–81.

Biederman, J., Rosenbaum, J. F., Bolduc-Murphy, E. A., et al., 'A 3-year follow-up of children with and without behavioral inhibition', *J. Am. Acad. Child Adolesc. Psychiatry* (1993), 32 (4): 814–21.

Cloninger, C. R., Svrakic, D. M., Przybeck, T. R., 'A psychobiological model of temperament and character', *Arch. Gen. Psychiatry* (1993), 50 (12): 975–90.

## Chapter 8: Gender and Hormones

## A Pregnant Pause

George, D. T., Ladenheim, J. A., Nutt, D. J., 'Effect of pregnancy on panic attacks', *Am. Journal Psychiatry* (1987), 144 (8): 1078–9.

Swales, P. J., Sheikh, J. I., 'Hysterectomy in patients with panic disorder' (letter), *Am. J. Psychiatry* (1992), 149: 846–7.

Dembert, M. L., Dinneen, M. P., Opsahi, M. S., 'Estrogen-induced panic disorder' (letter), *Am. J. Psychiatry* (1994), 151 (8): 1246.

Griez, E. J. L., Hauzer, R., Meijer, J., 'Pregnancy and estrogen-induced panic' (letter), *Am. J. Psychiatry* (1995), 152 (11): 1688.

## The Numbers Game

Cohen, L. S., Sichel, D. A., Dimmock, J. A., Rosenbaum, J. F., 'Impact of pregnancy on panic disorder: a case series', *J. Clinical Psychiatry* (1994), 55 (7): 284–8.

Cohen, L. S., Sichel, D. A., Dimmock, J. A., Rosenbaum, J. F., 'Postpartum course in women with preexisting panic', *J. Clinical Psychiatry* (1994), 55 (7): 289–92.

Klein, D. F., 'Pregnancy and panic disorder' (commentary), *J. Clinical Psychiatry* (1994), 55 (7): 293–4.

Cohen, L. S., Sichel, D. A., Faraone, S. V., et al., 'Course of panic disorder during pregnancy and the puerperium: a preliminary study', *Biol. Psychiatry* (1996), 39 (11): 950–54.

Villeponteaux, V. A., Lydiard, R. B., Laraia, M. T., Stuart, G. W., Ballenger, J. C., 'The effects of pregnancy on preexisting panic disorder', *J. Clinical Psychiatry* (1992), 53 (6): 201–3.

## Teamwork

Panay, N., Studd, J. W., 'The psychotherapeutic effects of oestrogens', *Gynecol. Endocrinol.* (1998), 12 (5): 353–65.

Yonkers, K. A., Halbreich, U., Freeman, E., et al., 'Symptomatic improvement of premenstrual dysphoric disorder with sertraline treatment: a randomized controlled trial', *JAMA* (1997), 278 (12): 983–8.

Asso, D., Beech, H. R., 'Susceptibility to the acquisition of a conditioned response in relation to the menstrual cycle', *Journal of Psychosomatic Research* (1975), 19: 337–44.

## Migraine and Marriage

Cooper, C., Dunbar, N., Mira, M., 'Sex and seasickness on the Coral Sea', *Lancet* (1997), 350: 892.

Sigvardsson, S., Bohman, M., von Knorring, A-L., Cloninger, C. R., 'Symptom patterns and causes of somatization in men: I differentiation of two discrete disorders', *Genetic Epidemiology* (1986), 3: 153–69.

Kushner, M. G., Sher, K. J., Beitman, B. D., 'The relation between alcohol problems and the anxiety disorders', *Am. J. Psychiatry* (1990), 147 (6): 685–95.

Cox, B. J., Swinson, R. P., Shulman, I. D., Kuch, K., Reichman, J. T., 'Gender effects and alcohol use in panic disorder with agoraphobia', *Behav. Res. Ther.* (1993), 31 (4): 413–16.

Nunes, E., Quitkin, F., Berman, C., 'Panic disorder and depression in female alcoholics', *J. Clin. Psychiatry* (1988), 49 (11): 441–3.

Hafner, R. J., 'The husbands of agoraphobic women: assortive mating or pathogenic interaction?', *British J. Psychiatry* (1977), 130: 233–9.

Monteiro, W., Marks, I. M., Ramm, E., 'Marital adjustment and treatment outcome in agoraphobia', *British J. Psychiatry* (1985), 146: 383–90.

Craske, M. G., Burton, T., Barlow, D. H., 'Relationships among measures of communication, marital satisfaction and exposure during couples treatment of agoraphobia', *Behav. Res. Ther.* (1989), 27 (2): 131–40.

Andrews, B., Brown, G. W., 'Stability and change in low self-esteem: the role of psychosocial factors', *Psychol. med.* (1995), 25 (1): 23–31.

## Girl Power

Nadelson, C. C., Notman, M. T., 'The impact of the new psychology of men and women on psychotherapy', in *American Psychiatric Press Review of Psychiatry*, vol. 10, A. Tasman and S. M. Goldfiger (eds.), Washington, DC, American Psychiatric Press (1991). ISBN: 0-88048-436-5.

Nadelson, C. C., 'Male–female relationships: 600 years of change', *The Psychoanalytic Review* (1986), 73 (4): 600–605.

## Chapter 9: Light and Electromagnetism

### Mary's Story

Robins, L., Helzer, J. E., Weissman, M. M., et al., 'Lifetime prevalence of specific psychiatric disorders in three sites', *Arch. Gen. Psychiatry* (1984), 41: 949–58.

### Magnets and Migration

Baker, R. R., Mather, J. G., Kenaugh, J. H., 'Magnetic bones in human sinuses' (letter), *Nature* (1983), 301: 78–80.

Baker, R. R., *Human Navigation and Magnetoreception*, Manchester University Press (1989). ISBN: 0-71902-627-X.

Baker, R. R., 'Human navigation and magnetoreception: the Manchester experiments do replicate', *Animal Behaviour* (1987), 35: 691–704.

Brodeur, P., *The Zapping of America*, New York, W. W. Norton (1977). ISBN: 0-39306-427-1.

Grön, G., Wunderlich, A. P., Spitzer, M., Tomczak, R., Riepe, M. W., 'Brain activation during human navigation: gender-different neural networks as substrate of performance', *Nature Neuroscience* (2000), 3 (4): 404–8.

### Strip Lights and Seizures

Watts, F. N., Wilkins, A. J., 'The role of provocative visual stimuli in agoraphobia', *Psychological Medicine* (1989), 19: 875–85.

Hazell, J., Wilkins, A. J., 'A contribution of fluorescent lighting to agoraphobia', *Psychological Medicine* (1900), 20: 591–6.

## Alternative Applications

Anderson, D. J., The treatment of migraine with variable frequency photo-stimulation', *Headache* (1989), 29: 154–5.

Anderson, D. J., Legg, N. J., Ridout, D. A., 'Preliminary trial of photic stimulation for premenstrual syndrome', *Journal of Obstetrics and Gynaecology* (1997), 17 (1): 76–9.

## Chapter 10: A Physical Problem?

### The Inner Ear

Levinson, H. N., *Phobia Free*, New York, M. Evans and Company Inc. (1986): ISBN: 0-87131-475-4.

Levinson, H. N., *A Scientific Watergate – Dyslexia: How and Why Countless Millions are Deprived of Breakthrough Medical Treatment*, New York, Stonebridge Publishing Ltd (1994): ISBN: 0-9639303-0-3.

Levinson, H. N., 'Abnormal optokinetic and perceptual span parameters in cerebellar and vestibular disfunction and related anxiety disorders' *Percept. Mot. Skills* (1989) 68: 471–84.

### Palpitations

Kantor, J. S., Zitrin, C. M., Zeldis, S. M., 'Mitral valve prolapse syndrome in agoraphobic patients', *Am. J. Psychiatry*, 137 (4) (1980): 467–9.

### Dyslexia and a Lack of Balance

Levinson, H. N., *Smart But Feeling Dumb*, New York, Warner Books Inc. (1984): ISBN: 0-44651-307-5.

Fawcett, A. J., Nicolson, R. I., 'Automisation deficits in balance for dyslexic children', *Percept. Mot. Skills* (1992), 75 (2): 507–29.

Nicolson, R. I., Fawcett, A. J., Dean, P., 'Time estimation deficits in developmental dyslexia: evidence of cerebellar involvement', *Proc. R. Soc. London B. Biol. Sci.* (1995), 259 (1354): 43–7.

Nicolson, R. I., Fawcett, A. J., Berry, E. L., et al., 'Association of abnormal cerebellar activation with motor learning difficulties in dyslexic adults', *Lancet* (1999), 353 (9165): 1662–7.

## Conclusion

### Into the Future

'Neuroscience 2000: a new era of discovery', symposium organised by the Society for Neuroscience, Washington, DC, 12–13 April 1999.

'Celebrating a decade of progress' (editorial), *Nature Neuroscience* (1999), 2 (6): 487.

Andreasen, N. C., 'Changing boundaries in psychiatry', *Lancet* (1999), 354: 56.

### The Next Hundred Years

Marks, I. M., Nesse, R. M., 'Fear and fitness: an evolutionary analysis of anxiety disorders', *Ethology and Sociobiology* (1994), 15: 247–61.

### Breaking Down Barriers

Morris, K., 'Advances in "brain decade" bring new challenges' (news), *Lancet* (2000), 355: 45.

Gabbard, G. O., MD, 'Psychodynamic psychiatry in the "Decade of the Brain"', *Am. J. Psychiatry* (1992), 149 (8): 991–8.

# Index